T0140335

Springer Theses

Recognizing Outstanding Ph.D. Research

Aims and Scope

The series "Springer Theses" brings together a selection of the very best Ph.D. theses from around the world and across the physical sciences. Nominated and endorsed by two recognized specialists, each published volume has been selected for its scientific excellence and the high impact of its contents for the pertinent field of research. For greater accessibility to non-specialists, the published versions include an extended introduction, as well as a foreword by the student's supervisor explaining the special relevance of the work for the field. As a whole, the series will provide a valuable resource both for newcomers to the research fields described, and for other scientists seeking detailed background information on special questions. Finally, it provides an accredited documentation of the valuable contributions made by today's younger generation of scientists.

Theses are accepted into the series by invited nomination only and must fulfill all of the following criteria

- They must be written in good English.
- The topic should fall within the confines of Chemistry, Physics, Earth Sciences, Engineering and related interdisciplinary fields such as Materials, Nanoscience, Chemical Engineering, Complex Systems and Biophysics.
- The work reported in the thesis must represent a significant scientific advance.
- If the thesis includes previously published material, permission to reproduce this must be gained from the respective copyright holder.
- They must have been examined and passed during the 12 months prior to nomination.
- Each thesis should include a foreword by the supervisor outlining the significance of its content.
- The theses should have a clearly defined structure including an introduction accessible to scientists not expert in that particular field.

More information about this series at http://www.springer.com/series/8790

Donato Zarrilli

Integration of Low Carbon Technologies in Smart Grids

Doctoral Thesis accepted by
the University of Siena, Siena, Italy

Author
Dr. Donato Zarrilli
Dipartimento di Ingegneria dell'Informazione e Scienze Matematiche
Università di Siena
Siena, Italy

Supervisor
Prof. Antonio Vicino
Dipartimento di Ingegneria dell'Informazione e Scienze Matematiche
Università di Siena
Siena, Italy

ISSN 2190-5053 ISSN 2190-5061 (electronic)
Springer Theses
ISBN 978-3-030-07487-6 ISBN 978-3-319-98358-5 (eBook)
https://doi.org/10.1007/978-3-319-98358-5

This Springer imprint is published by the registered company Springer Nature Switzerland AG
The registered company address is: Gewerbestrasse 11, 6330 Cham, Switzerland

To my father Giuseppe.
In memoriam.

To my mother Maria Teresa
and brother Valentino.
With love.

Supervisor's Foreword

The ever-increasing penetration of low carbon technologies in electricity systems is opening new challenging research avenues on integration of renewable energy sources in the smart grid. Donato Zarrilli obtained important novel results in this really exciting field.

The major contributions of his Ph.D. dissertation fall in the following research areas:

1. Market and physical issues related to the integration of renewables in the power system chain;
2. Smart home and building HVAC control systems for coordination of distributed renewable energy sources and demand flexibility (demand response);
3. Role of energy storage systems in the electricity grids of the future: decision strategies for placement, sizing, and operation of storage devices in distribution networks.

With reference to scientific achievements obtained by Donato, I would like to stress the solution of the fundamental decision problem on how to allocate energy storage systems in a distribution network, in the presence of uncertainty due to stochastic renewable generation and demand. Moreover, he provided important contributions on receding horizon control techniques for robust and reliable operation of storage devices in active distribution networks.

The second important achievement consists in embedding demand response services in a model predictive control framework for optimal management of HVAC systems in large-scale buildings.

The third main contribution regards optimal bidding of wind energy on the electricity market. As a matter of fact, in view of their inaccuracies, wind speed forecasts cannot be used directly to formulate bids for the day-ahead market. Nevertheless, they can be fruitfully exploited by embedding them in a classification framework which enhances consistently the performance of the resulting bidding strategy with respect to alternative techniques available in the literature.

The contribution of Donato to the literature on the research topics of the thesis has been really substantial, from both the qualitative point of view and the quantitative point of view. Actually, he has devised sophisticated models as well as control and optimization techniques, testing them successfully on real data. For these reasons, the thesis of Dr. Zarrilli was awarded the Prize as the Best Ph.D. thesis in the period June 2016–May 2017 by the Italian Society for Research in Systems and Control Engineering (SIDRA).

Donato has carried out his research in the Systems and Control Lab of the University of Siena, under my supervision and the co-advisorship of Antonio Giannitrapani and Simone Paoletti.

I would like to close this foreword by recognizing Donato's research talent and implementation capabilities, as well as his exceptional interpersonal skills.

Siena, Italy
April 2018

Prof. Antonio Vicino

Parts of this thesis have been published in the following articles:

- D. Zarrilli, A. Giannitrapani, S. Paoletti, A. Vicino, *Energy storage operation for voltage control in distribution networks: a receding horizon approach*. IEEE Transactions on Control Systems Technology, vol. 26, no. 2, pp. 599–609, 2018.
- A. Giannitrapani, S. Paoletti, A. Vicino, D. Zarrilli, *Optimal allocation of energy storage systems for voltage control in LV distribution networks*. IEEE Transactions on Smart Grids, vol. 8, no. 6, pp. 2859–2870, 2017.
- G. Bianchini, M. Casini, A. Vicino, D. Zarrilli, *Demand-response in building heating systems: a model predictive control approach*. Applied Energy, vol. 168, pp. 159–170, 2016.
- A. Giannitrapani, S. Paoletti, A. Vicino, D. Zarrilli, *Bidding wind energy exploiting wind speed forecasts*. IEEE Transactions on Power Systems, vol. 31, no. 4, pp. 2647–2656, 2016.

Acknowledgements

First and foremost I want to thank my advisor Prof. Antonio Vicino. It has been an honor to be his Ph.D. student. I appreciate all his contributions of time, ideas, and funding to make my doctoral experience productive and stimulating. The joy and enthusiasm he has for the research was contagious and motivational for me, even during tough times. I am also thankful for the excellent example he has provided both as a scientist and successful man.

The members of the Control Systems Group have contributed immensely to my personal and professional growth. The group has been a source of friendships as well as good advice and collaboration. A special thank goes to Drs. Antonello Giannitrapani and Simone Paoletti for their invaluable guide throughout these years. I also have had the pleasure to work with Drs. Gianni Bianchini and Marco Casini, postdocs Alessandro and Toufik, and the numerous students who have come through the laboratory.

I wish to extend my gratitude to Prof. Pierluigi Mancarella and to the people at the Electrical Energy and Power Systems Group, for the way they have welcomed me and for the pleasant time we spent together at the University of Manchester. I also thank the committee for their willingness to attend my thesis defense, especially Prof. Arturo Losi.

My time in Siena was made enjoyable in large part due to the many friends that became a part of my life. I am grateful for time spent with flatmates Alessio, Angelo, and Ennio, and for many other people and memories.

Last but not least, importantly, I would like to thank my family for all their love and encouragement. In particular, my deepest thanks go to my parents Giuseppe and Maria Teresa, for their patience and support in all my pursuits. I love you.

Contents

Acronyms

AC	Alternating current
ARMA	Autoregressive moving average
ARX	Autoregressive exogenous
cdf	Cumulative distribution function
CSA	Cluster and sensitivity analysis
DC	Direct current
DG	Distributed generation
DR	Demand response
DSO	Distribution system operator
ES	Exponential smoothing
ESS	Energy storage system
HVAC	Heating, ventilating, and air-conditioning system
LP	Linear program
LV	Low voltage
MILP	Mixed-integer linear program
MPC	Model predictive control
MRLP	Multicategory robust linear programming
MSE	Mean square error
MV	Medium voltage
OLTC	On-load tap changer
OPF	Optimal power flow
pdf	Probability density function
PRBS	Pseudo-random binary sequence
PV	Photovoltaic
RES	Renewable energy source
SDP	Semidefinite programming
UC	Unit commitment
WPP	Wind power producer

Symbols, Operators, and Functions

$\mathbb{R}$	Field of real numbers
$\mathbb{N}$	Set of natural numbers
$\mathbb{C}$	Field of complex numbers
$\mathbb{B} = \{0, 1\}$	Binary set
$\in$	Belongs to
$x \in \mathbb{R}$	Real number
$x \in \mathbb{C}$	Complex number
$\widehat{x}$	Estimate of x
$\boldsymbol{x} \in \mathbb{R}^{n\times m}$	Real $n \times m$ matrix
$\boldsymbol{x} \in \mathbb{C}^{n\times m}$	Complex $n \times m$ matrix
$\jmath$	Imaginary unit
$\mathrm{Pr}(\cdot)$	Probability of the underlying event
$\mathrm{Re}(\cdot)$, $\mathrm{Im}(\cdot)$	Real and imaginary parts of a complex number
$\mathrm{Tr}(\cdot)$	Trace operator
$(\cdot)^*$	Complex conjugate transpose operator
$(\cdot)'$	Transpose operator
$\lvert \cdot \rvert$	Absolute value operator
$\mathbf{E}[\cdot]$	Expectation operator
max, min	Maximum or minimum operator
arg max, arg min	Maximizing or minimizing argument
$\succeq, \preceq$	Inequality signs in the positive semidefinite sense
$\subseteq$	Subset of
$\mathcal{X} \backslash \{x\}$	Set $\mathcal{X}$ without the element x

Chapter 1
Introduction

Electricity is all around us. Whether we are at home or in the workplace, in a busy metropolis or in a remote outpost, electricity, whether directly or indirectly, enables almost every-thing we do. Electricity is in many respects an ideal means of transmitting and delivering energy, being controllable, safe, economic, efficient and relatively unobtrusive.

Claes Rytoft,
ABB Group Senior Vice President

1.1 Background

The entire electricity system is undergoing a dramatic transformation driven mainly by challenging environmental and economic targets set out by government policies worldwide [1–3]. The general consensus about the necessary changes towards clean energies has fostered growth in renewable generation through different incentives programs. The restructuring process aims at "*decarbonizing*" the electricity sector while increasing requirements in terms of quality of supply and making electricity more affordable to end customers. In contrast to traditional power systems, in which a small number of centralized plants connected at the transmission level supplied the surrounding consumption areas, today we are witnessing a massive proliferation of renewable generation at the distribution side. Since renewable power plants, such as solar and wind, are typically distributed over many locations and even integrated in consumers' premises, a given node can arbitrarily change from being a passive point to an active one. In this way, the standard model of one-way (downstream) power flows, where generators are controlled to follow demand levels, is giving way to multiple and more complex interactions resulting in bi-directional (both downstream and upstream) flows [4, 5]. However, the transition from passive to active networks seems to follow the "*installing*" rather than the "*integrating*" policy. This massive and unco-

D. Zarrilli, *Integration of Low Carbon Technologies in Smart Grids*, Springer Theses, https://doi.org/10.1007/978-3-319-98358-5_1

ordinated penetration of such distributed energy resources is completely changing the way the electricity grid works and it is bringing serious problem to network operators. In fact, the intrinsic variability of renewable generation is posing new stress to system balancing, with the risk of curtailing clean energy at times of system congestions [6, 7]. As a consequence, prices in day-ahead and especially balancing markets are becoming more volatile, and renewable power producers are being called to take part in the intrinsic cost of intermittent production. This may hinder the benefits introduced by renewables, reduce the overall efficiency as well as the expectations of cheaper electricity prices. It is straightforward that these changes are affecting the whole electricity system, from the hardware of the network infrastructure to the way it is operated. In this light, to meet this challenge, a new technical architecture along with different market mechanisms is required in the future scenario that will enable smart grids to accommodate the intermittency and the relevant uncertainty of renewable generation and become the means for cost efficient and sustainable energy supply systems [8, 9]. For example, the energy network balancing can no longer rely on supply strictly following demand approach, but must be accomplished by involving both generation and demand. In recent years, several methodologies that allow active participation of consumers in power system operation have drawn more and more attention [10]. However, this requires advanced monitoring, communication and control systems across the entire power system chain.

1.2 Thesis Contribution and Organization

Power system operators are constantly under pressure to reduce line losses and peak demand, better support feeder voltage and power flows, and increase efficiency and power quality to end consumers. Furthermore, the cost of traditional grid reinforcement, as well as environmental considerations, has driven them to find more effective ways to meet future network requirements. In this context, the thesis provides different opportunities and ideas to help face some of these challenges. In particular the work is focused on the effective integration of distributed low carbon technologies in the grid of the future. Planning and operation problems for different clean solutions, such as market bidding strategies for intermittent energy producers, demand side management algorithms for smart buildings, and electrical storage options for network operators, have been studied for facilitating the integration of renewable energy sources (RES) in the power network.

Most relevant decisions to be made by network agents within a fast-moving energy environment involve copying with large quantities of data and a significant level of uncertainty. For example, future renewable electricity production is unknown when producers have to submit their offers to the market. In a similar fashion, at the time of procuring energy to be supplied, retailers do not know their consumers' demand. In this respect, it is fundamental to properly address the uncertainty involved and exploit all the available information. Additionally, large and complex optimization problems, which are generally prone to numerical issues or simply take too long to converge, are needed to be solved to make informed and economic decisions. In fact, more often than not, planning and control problems in power systems are formulated as mixed-integer non-linear programs involving a huge number of optimization variables and physical constraints. It would be enough to mention the decision-making process required for the optimal configuration of energy storage devices in the electrical network. The full detailed formulation is composed of an AC (non-linear) optimal power flow (OPF) with the addition of integer variables. In such a situation, ad-hoc simplifications would be necessary when approaching realistic applications, while maintaining the accuracy of the solution at an acceptable degree.

The interdisciplinary research presented in the next chapters lies at the intersection of power systems, optimization techniques and controls. This work further provides a number of optimization tools employing stochastic models and different physical-based heuristics that are amenable to fast and robust computation. To that effect, appropriate linear or convex formulations are embedded into most of the proposed models, which involve market/system operators, renewable producers and consumers. Particular attention is paid to electric power systems at large extent, including spot electricity markets and smart buildings, with a large integration of non-dispatchable sources, such as wind and solar power plants, and a strong need for demand energy management. The thesis is organized as follows.

Chapter 2 addresses the problem of determining the optimal day-ahead bidding strategy for a wind power producer (WPP) by exploiting wind speed forecasts provided by a meteorological service. In the considered framework, WPP is called to take part in the responsibility of system operation by providing day-ahead generation profiles and undergoing penalties in case of deviations from the schedule. Penalties are applied only if the delivered hourly energy deviates from the schedule more than a given relative tolerance. The optimal solution is obtained analytically by formulating and solving a stochastic optimization problem aiming at maximizing the expected profit. The proposed approach consists in exploiting wind speed forecasts to classify the next day into one of several predetermined classes, and then selecting the optimal solution derived for each class.

Chapter 3 deals with the problem of optimizing the operation of a building heating system under the hypothesis that the building is included as an active consumer in a demand response (DR) program. DR requests to the building operational system come from an external market player or a grid operator. Requests assume the form of price-volume signals specifying a maximum volume of energy to be consumed during a given time slot and a monetary reward assigned to the participant in case the conditions are fulfilled. A receding horizon control approach is adopted for the minimization of the energy bill, by exploiting a simplified model of the building. Since the resulting optimization problem is a mixed integer linear program which turns out to be manageable only for buildings with very few zones, a heuristic is devised to make the technique applicable to problems of practical size.

Chapter 4 discusses the role of energy storage system (ESS) in preventing over- and undervoltages in low voltage (LV) distribution networks. Both the problems of finding the optimal configuration (number, location and size) and computing the real-time control policy of such distributed devices, are presented. A heuristic strategy based on the network voltage sensitivity analysis is proposed to identify the most effective buses where to install a given number of ESSs, while circumventing the combinatorial nature of the problem. For fixed storage locations, the multi-period OPF framework is adopted to formulate the sizing problem, for whose solution convex relaxations based on semidefinite programming (SDP) are exploited. Uncertainties in the storage sizing decision problem due to stochastic generation and demand, are accounted for via a scenario approach which considers different realizations of the demand and generation profiles. The final choice of the most suitable storage allocation is done by minimizing a total configuration cost, which takes into account the number of storage devices, their total installed capacity and average network losses. A voltage control scheme based on a receding horizon approach to optimally operate the ESSs installed in the network is also developed. The essential feature of the proposed control approach lies in the very limited information needed to predict possible voltage problems, and to compute the ESS control policy making it possible to counteract them in advance.

Chapter 5 draws a summary of the thesis contribution and a presents a brief discussion of the achieved results and future research directions.

References

1. Beecher JA, Kalmbach JA (2012) Climate change and energy. U.S, National Climate Assessment Midwest Technical Input Report
2. International Energy Agency (2015) Energy and climate change. World Energy Outlook Special Report
3. Kasemir B, Dahinden U, Swartling ÅG, Schüle R, Tabara D, Jaeger CC (2000) Citizens perspectives on climate change and energy use. Glob Environ Change 10(3):169–184
4. Pepermans G, Driesen J, Haeseldonckx D, Belmans R, Haeseleer WD (2005) Distributed generation: definition, benefits and issues. Energy Policy 33(6):787–798
5. Willis HL (2000) Distributed power generation: planning and evaluation. CRC Press
6. Boyle G (2012) Renewable electricity and the grid: the challenge of variability. Earthscan
7. Phuangpornpitak N, Tia S (2013) Opportunities and challenges of integrating renewable energy in smart grid system. Energy Proced 34:282–290
8. Farhangi H (2010) The path of the smart grid. IEEE Power Energ Mag 8(1):18–28
9. Ipakchi A, Albuyeh F (2009) Grid of the future. IEEE Power Energ Mag 7(2):52–62
10. Losi A, Mancarella P, Vicino A (2015) Integration of demand response into the electricity chain: challenges, opportunities and smart grid solutions. Wiley

Chapter 2
Bidding Renewable Energy in the Electricity Market

In this chapter the problem of designing optimal bidding strategies for renewable power producers exploiting weather forecasts is addressed. The material of this chapter is mainly based on [1, 2].

2.1 Introduction

In recent years, the interest in generating power from RESs has grown rapidly, pushed by the expected benefits both in environmental and economic terms [3]. On the other hand, due to their intrinsic intermittency and variability, RES integration in the grid is causing serious problems to transmission and distribution system operators [4]. One possible way to mitigate the uncertainty of RES generation is to require that producers provide day-ahead generation profiles, and to apply penalties if the delivered energy over a specified time period deviates from the schedule more than a given relative tolerance. This mechanism involves RES producers in the intrinsic risk of intermittent production and calls for suitable strategies enabling them to maximize their expected profit in front of generation uncertainty. Indeed, depending on the structure of the penalties, the strategy which maximizes the expected profit for the producer could not correspond to merely offering the generation forecasts obtained with the objective to minimize the forecasting errors. In this chapter, the above problem for the case of WPPs is addressed. It is assumed that a WPP makes bids on a day-ahead market, and these bids represent a delivery obligation for it. Therefore, the problem of optimizing the day-ahead generation profiles can be seen as the problem of optimizing the bids made on the market. The problem of designing optimal WPP bidding strategies has been addressed in a considerable amount of papers in the last decade. Several contributions focus on advanced techniques for obtaining highly reliable wind point predictions, and relate the reliability of the forecasts to the regulation cost of imbalances (see, e.g., [5, 6]). Other contributions, such as [7–10],

D. Zarrilli, *Integration of Low Carbon Technologies in Smart Grids*,
Springer Theses, https://doi.org/10.1007/978-3-319-98358-5_2

embed the bidding strategy within a stochastic framework, where the sources of randomness, i.e., wind variability and market prices, are blended together. Then, a stochastic optimization approach is adopted to derive optimal contract volumes for the WPP. An excellent overview of these different approaches is provided in [8], including a comparison between techniques using wind point forecasts and those using the wind power probability density function (*pdf*) of the generation plant. In [9, 10] the stochastic approach is analyzed in the two distinct cases in which the imbalance price statistics is assumed independent or not of the weather statistics. Optimal bids are derived in [10] under the assumption of statistical independence, and the role played by a possibly co-located thermal generator or storage device is analyzed. In [9] a general approach is taken by explicitly considering statistical dependence of wind and imbalance prices. Alternative approaches, such as [11–13], require the knowledge of a large number of possible scenarios related to wind outturn uncertainty. In [12, 14], either a risk sensitive term is introduced in the cost function to model WPP risk-averse preferences (VaR or CVaR), or a utility function is adopted to suitably shape the cost function. The role of an ESS in increasing reliability of bids and mitigating the financial risks of the WPP has been investigated in [15]. Finally, a number of papers deals with other problems relevant to wind power bidding strategies, such as coordination strategies of wind generation and reserve or storage (see, e.g., [16] and references therein), cost of wind generation forecast errors [17] and advanced wind power forecasting [18, 19].

The present contribution can be positioned in the line of research adopting stochastic models for optimizing the bids of wind power in a day-ahead market. In this context, the work exploits the basic optimal bidding strategy introduced in [8, 10], with the aim of extending the results in two directions. The first direction is to derive analytically the optimal bidding strategy for the case with penalties applied only outside tolerance bands around the nominal bidding profile. The solution is based on the knowledge of the prior wind energy cumulative distribution function (*cdf*) of the generation plant. The second direction consists in investigating the use of weather forecasts to further enhance the bidding strategy. The idea is to use wind speed forecasts to classify the next day (e.g., windy or windless day), and then to replace the wind energy unconditional *cdf* with the "conditional" *cdf* of the selected class. It is worthwhile to notice that the introduction of the classifier is in many respects similar to the adoption of scenarios in the complicated setting where the dependence of wind and imbalance prices cannot be neglected. In the proposed approach, the problem is simplified by adopting a family of possible scenarios parameterized according to the structure of the classifier. In turn, the classifier parameters are inferred from historical data. As a further contribution of this chapter, a parametric model for the probability distribution of the energy generated by a wind power plant is proposed. The parametric model is obtained by mixing a beta distribution and a truncated gamma distribution. Bidding strategies for solar photovoltaic (PV) power producers, dealing with non-stationarity of solar PV energy distribution, are proposed in [20, 21].

The chapter is organized as follows. In Sect. 2.2 the considered bidding problem is formulated and its optimal solution is derived. Section 2.3 proposes a parametric model for the wind energy distribution. The use of wind speed forecasts to enhance the performance of the bidding strategy is investigated in Sect. 2.4. Section 2.5 reports experimental results and, finally, conclusions are drawn in Sect. 2.6.

2.2 Optimal Bidding Strategy

The following market setup, which extends the ones in [8, 10], is considered. A WPP participates in a spot market (for example, a day-ahead market) where it offers ex-ante the hourly energy blocks b_t [kWh], $t = 1, \dots, 24$. The WPP is assumed to have zero marginal cost of production and to be a price taker in the spot market. This is motivated by the fact that the individual WPP capacity is negligible relative to the whole market [10]. For the sake of simplicity, it is assumed that the bid b_t is not curtailed at market closure, and represents a delivery obligation for the WPP. Let e_t [kWh] be the energy actually generated by the wind power plant during the hour t of the day. The WPP is subject to *ex-post* financial penalties for deviations of e_t from b_t which exceed predefined thresholds. In this section, the problem of optimizing the bids b_t in the above scenario is formulated. The optimal solution is then derived in terms of the wind energy statistics and the imbalance penalties. Note that these penalties depend on the considered regulation mechanism. For instance, they could be a given fraction of the market clearing price, or follow a more complex function [8]. However, the presented results and analysis do not rely on any specific assumption about the regulation mechanism.

Assume that λ_t [€/kWh] is the clearing price in the spot market for the hour t of the day. Given $\sigma \in [0, 1]$, this price is used to fully reward the portion of e_t which does not exceed the upper bound $(1 + \sigma)b_t$ above the nominal contract b_t. The relative tolerance σ is assumed to be fixed by existing regulation. In case e_t is smaller than $(1 - \sigma)b_t$, the undelivered quantity $(1 - \sigma)b_t - e_t$ is penalized with price λ_t^- [€/kWh]. Vice versa, if e_t is greater than $(1 + \sigma)b_t$, the surplus quantity $e_t - (1 + \sigma)b_t$ is remunerated with price $\lambda_t - \lambda_t^+$, where λ_t^+ [€/kWh] is the penalty applied with respect to the settlement price λ_t. It follows that the net hourly profit for the WPP amounts to

$$\begin{aligned} \Pi(b_t, e_t) = {} & \lambda_t e_t - \lambda_t^- \max\{(1 - \sigma)b_t - e_t, 0\} \\ & - \lambda_t^+ \max\{e_t - (1 + \sigma)b_t, 0\}. \end{aligned} \tag{2.1}$$

Remark 1 Though the market mechanism described so far introduces penalties in case of deviations from the offered contract ($\lambda_t^- \geq 0$, $\lambda_t^+ \geq 0$), the formulation (2.1) also fits to the case when deviations are actually rewarded. This is obtained by setting $\lambda_t^- < 0$ or $\lambda_t^+ < 0$. For instance, a regulatory framework introduced in Italy at the beginning of 2013 assigned a reward whenever the WPP imbalance was of the opposite sign with respect to the imbalance at regional level. In this case, even though the WPP did not comply with its commitment, it was nevertheless rewarded because it actually helped the system.

Remark 2 Notice that, if $0 \leq \lambda_t^+ \leq \lambda_t$, the surplus quantity $e_t - (1+\sigma)b_t$ is still remunerated with nonnegative price. That is, the WPP still gains from delivering the surplus energy. On the other hand, if $\lambda_t^+ > \lambda_t$, delivery of the energy surplus actually decreases the net profit for the WPP. In practice, such a scenario could be avoided provided that the WPP has curtailment capabilities.

Remark 3 The profit in (2.1) can be seen as a generalization of that considered in [10, 22], which takes the form

$$\begin{aligned} \widetilde{\Pi}(b_t, e_t) = \ & \lambda_t b_t - q_t \max\{b_t - e_t, 0\} \\ & - p_t \max\{e_t - b_t, 0\}. \end{aligned} \tag{2.2}$$

Indeed, when $\sigma = 0$, i.e., there is no tolerance interval around the nominal bid, the profits in (2.1) and (2.2) coincide by selecting $\lambda_t^- = q_t - \lambda_t$ and $\lambda_t^+ = p_t + \lambda_t$.

The profit in (2.1) is a stochastic quantity due to the uncertainty on the generated energy e_t. Therefore, the considered optimal bidding problem consists in determining the bid b_t^* maximizing the expected profit $J(b_t) = \mathbf{E}[\Pi(b_t, e_t)]$, i.e.,

$$b_t^* = \arg \max_{b_t \in [0, \overline{e}]} J(b_t), \tag{2.3}$$

where $\mathbf{E}[\cdot]$ denotes expectation with respect to the probability measure of e_t, and $\overline{e}$ [kWh] is the maximum amount of energy that the wind power plant can produce in one hour.

The solution to the optimization problem (2.3) is derived next. To this aim, let $F_t(\xi) = \Pr(e_t \leq \xi)$ be the *cdf* of the random variable e_t, and $f_t(\xi)$ be the corresponding *pdf*. In the following, it is assumed that $F_t(\xi)$ is differentiable, so that $f_t(\xi)$ exists for all $\xi \in (0, \overline{e})$. In order to characterize the optimal bid b_t^*, the following partition of $\mathbb{R}^2$, which is shown in Fig. 2.1, is introduced:

$$\begin{aligned} \mathfrak{R}_1 &= \left\{(\lambda^-, \lambda^+) \in \mathbb{R}^2 : \lambda^+ \leq 0,\ \mu_t \lambda^+ - g_t(\sigma)\lambda^- \leq 0\right\} \\ \mathfrak{R}_2 &= \left\{(\lambda^-, \lambda^+) \in \mathbb{R}^2 : \lambda^- > 0,\ \lambda^+ > 0\right\} \\ \mathfrak{R}_3 &= \left\{(\lambda^-, \lambda^+) \in \mathbb{R}^2 : \lambda^- \leq 0,\ \mu_t \lambda^+ - g_t(\sigma)\lambda^- > 0\right\}, \end{aligned} \tag{2.4}$$

where $\mu_t = \mathbf{E}[e_t]$ and

$$g_t(\sigma) = \int_0^{(1-\sigma)\overline{e}} ((1-\sigma)\overline{e} - \xi) f_t(\xi) d\xi. \tag{2.5}$$

Proposition 1 *The solution to (2.3) is*

$$b_t^* = \begin{cases} 0 & \text{if } (\lambda_t^-, \lambda_t^+) \in \mathfrak{R}_1 \\ \tilde{b}_t^* & \text{if } (\lambda_t^-, \lambda_t^+) \in \mathfrak{R}_2 \\ \overline{e} & \text{if } (\lambda_t^-, \lambda_t^+) \in \mathfrak{R}_3, \end{cases} \tag{2.6}$$

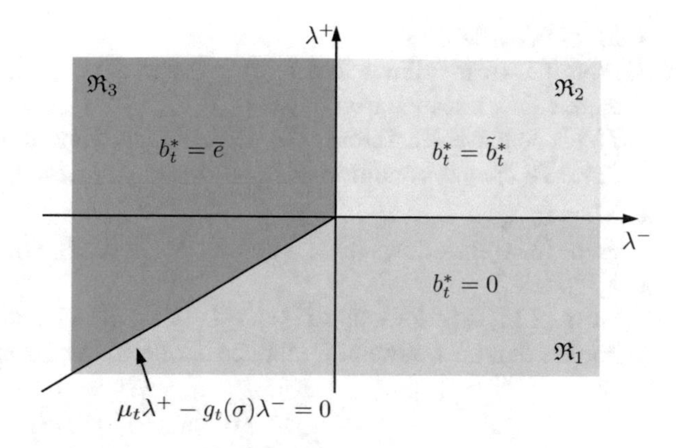

Fig. 2.1 Partition of the (λ^-, λ^+)-plane and optimal solution of problem (2.3) in each region of the partition

where $\mathfrak{R}_1$, $\mathfrak{R}_2$ *and* $\mathfrak{R}_3$ *are defined in (2.4)–(2.5) and* $\tilde{b}_t^*$ *satisfies the equation*

$$(1 - \varphi_t)(1-\sigma)F_t\big((1-\sigma)\tilde{b}_t^*\big) = \varphi_t(1+\sigma)\Big(1 - F_t\big((1+\sigma)\tilde{b}_t^*\big)\Big) \tag{2.7}$$

with φ_t *given by*

$$\varphi_t = \frac{\lambda_t^+}{\lambda_t^- + \lambda_t^+}. \tag{2.8}$$

Proof from the definition of $J(b_t)$ one gets

$$\begin{aligned} J(b_t) = \lambda_t\mu_t - \lambda_t^- \int_0^{(1-\sigma)b_t} ((1-\sigma)b_t - \xi)f_t(\xi)d\xi \\ - \lambda_t^+ \int_{(1+\sigma)b_t}^{\bar{e}} (\xi - (1+\sigma)b_t)f_t(\xi)d\xi. \end{aligned} \tag{2.9}$$

Under regularity assumptions on the *pdf* $f_t(\xi)$, the application of the Leibniz integral rule yields

$$\begin{aligned} J'(b_t) = &- \lambda_t^-(1-\sigma)F_t\left((1-\sigma)b_t\right) \\ &+ \lambda_t^+(1+\sigma)\big(1 - F_t\left((1+\sigma)b_t\right)\big). \end{aligned} \tag{2.10}$$

Since $F_t(\cdot)$ is differentiable, $J'(\cdot)$ is a continuous function. Besides, the second derivative of $J(\cdot)$ takes on the form

$$\begin{aligned} J''(b_t) = &- \lambda_t^-(1-\sigma)^2 f_t\left((1-\sigma)b_t^*\right) \\ &- \lambda_t^+(1+\sigma)^2 f_t\left((1+\sigma)b_t^*\right). \end{aligned} \tag{2.11}$$

In the following, the optimal solution is derived separately for each quadrant of the (λ_t^-, λ^+)-plane.

- $\lambda_t^- > 0, \lambda_t^+ > 0$.
  From (2.10) it follows that $J'(0) \geq 0$ and $J'(\overline{e}) \leq 0$. Since $J'(\cdot)$ is continuous, there exists a stationary point $\bar{x} \in [0, \overline{e}]$, i.e., $J'(\bar{x}) = 0$. Moreover, from (2.11), $J(\cdot)$ is concave. Therefore, $J(\cdot)$ attains its maximum over $[0, \overline{e}]$ at $\bar{x}$, i.e., $b_t^* = \bar{x}$. Since $J'(\bar{x}) = 0$, condition (2.7) follows easily from (2.10) and $\lambda_t^+ = \frac{\varphi_t}{1-\varphi_t}\lambda_t^-$.
- $\lambda_t^- \leq 0, \lambda_t^+ > 0$.
  From (2.10) it follows that $J'(x) \geq 0, \ \forall x \in [0, \overline{e}]$, and hence $b_t^* = \overline{e}$.
- $\lambda_t^- \leq 0, \lambda_t^+ \leq 0$.
  From (2.11) it follows that $J''(x) \geq 0, \ \forall x \in [0, \overline{e}]$, and hence $J(\cdot)$ is convex. This implies that the optimum is attained at one of the extrema of the feasible interval, i.e.,

$$b_t^* = \begin{cases} 0 & \text{if } J(0) \geq J(\overline{e}) \\ \overline{e} & \text{otherwise.} \end{cases} \tag{2.12}$$

  From (2.9), one gets $J(0) = (\lambda_t - \lambda_t^+)\mu_t$ and $J(\overline{e}) = \lambda_t\mu_t - g_t(\sigma)\lambda_t^-$, and hence

$$b_t^* = \begin{cases} 0 & \text{if } \mu_t\lambda_t^+ - g_t(\sigma)\lambda_t^- \leq 0, \\ \overline{e} & \text{otherwise.} \end{cases} \tag{2.13}$$

  Notice that, from (2.5), $g_t(\sigma) \geq 0$.
- $\lambda_t^- > 0, \lambda_t^+ \leq 0$.
  In this case, from (2.10) it follows that $J'(x) \leq 0, \ \forall x \in [0, \overline{e}]$, and hence $b_t^* = 0$.

The thesis follows from the expression of the partition (2.4). ■

Given the monotonicity of the first derivative of $J(b_t)$ for $(\lambda_t^-, \lambda_t^+) \in \mathfrak{R}_2$, numerical computation of the optimal contract $\tilde{b}_t^*$ in (2.7) can be performed very efficiently through bisection. Note that the optimal bidding strategy (2.6) depends on the knowledge of the pair $(\lambda_t^-, \lambda_t^+)$. If λ_t^- and λ_t^+ are stochastic variables independent of e_t, Proposition 1 still holds by replacing λ_t^- and λ_t^+ in (2.6) with the corresponding expected values. In practice, if λ_t^- and λ_t^+ are not known beforehand, one should replace them with suitable forecasts.

The following corollary of Proposition 1 establishes the connection between the presented result for $\sigma = 0$ and the analogous result found in [10] (see Remark 3).

Corollary 1 *When $\sigma = 0$, the optimality condition (2.7) boils down to $F_t(\tilde{b}_t^*) = \varphi_t$.*

2.3 Parametric Models of Wind Energy Distribution

An important issue arising from Proposition 1 regards the knowledge of the *cdf* $F_t(\cdot)$, or the *pdf* $f_t(\cdot)$, of the generated energy e_t, to be used for the computation of the optimal bid b_t^* when $(\lambda_t^-, \lambda_t^+) \in \mathfrak{R}_2$. In real applications, these functions are typically estimated from historical data. Computing the empirical *cdf*s is the simplest choice.

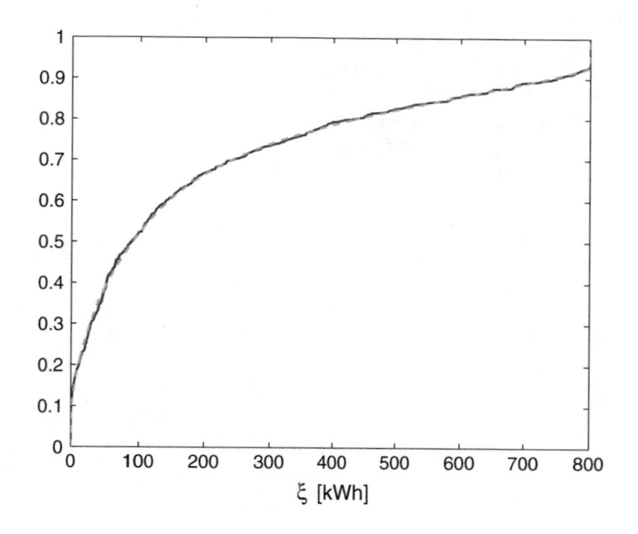

Fig. 2.2 Empirical *cdf* (solid) of the random variable e_{20} for a 800 kW wind turbine and corresponding BGM model (dashed) estimated from data. The two curves are in practice overlapping

Nevertheless, a parametric model for the wind energy statistics is more attractive and desirable for several reasons. The main motivation consists in the compression of the information in few parameters. Differences in the parameters may enable a better understanding of the wind energy behavior in different time frames, for instance by characterizing the statistics in terms of stationarity or cyclostationarity of the underlying stochastic process. Moreover, analysis of the evolution of the empirical *cdf* s with time clearly shows the presence of non-stationary features. In this respect, a parametric model allows one to exploit any of the large variety of parameter tracking techniques to follow the process changes.

A common feature emerging from the analysis of generation data sets from different wind plants, is that the *cdf* of the hourly generated energy e_t is discontinuous at $\xi = 0$ and $\xi = \overline{e}$ (see Fig. 2.2 for an example of empirical *cdf*). Discontinuity at $\xi = 0$ is motivated by the fact that wind turbines do not generate usable power below the cut-in and above the cut-out wind speed, while discontinuity at $\xi = \overline{e}$ is present because wind turbines generate nominal power between rated and cut-out wind speed (see Fig. 2.3 for a typical energy curve of a wind turbine). In view of this problem, a statistical model containing both discrete and continuous components turns out to be appropriate to describe the *pdf* of the process:

$$f(\xi) = f^D(\xi) + f^C(\xi). \tag{2.14}$$

The discrete component $f^D(\xi)$ is the sum of two scaled and shifted Dirac delta functions at discontinuity points $\xi = 0$ and $\xi = \overline{e}$, i.e.,

$$f^D(\xi) = \kappa_0\delta(\xi) + \kappa_1\delta(\xi - \overline{e}), \tag{2.15}$$

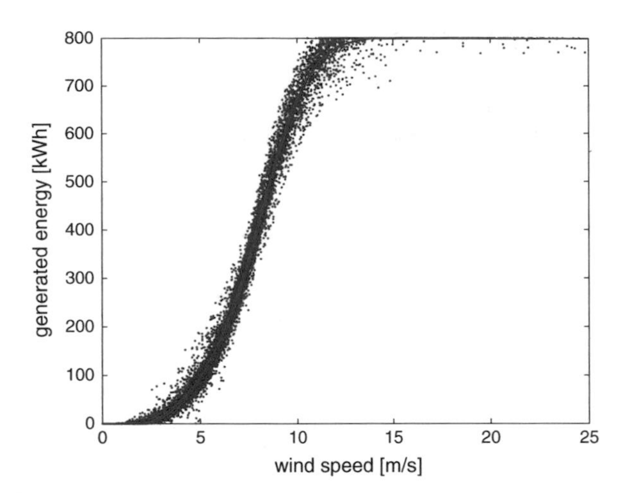

Fig. 2.3 Hourly generated energy versus hourly average wind speed for a 800 kW wind turbine (red points), and model (2.21) fitted to the data (solid blue curve)

where κ_0 and κ_1 are probability masses, while $f^C(\xi)$ is a continuous nonnegative function with support $(0, \overline{e})$ such that $\int_0^{\overline{e}} f^C(\xi)d\xi = 1 - \kappa_0 - \kappa_1$. Given a data set for the random variable e_t, parameters κ_0 and κ_1 can be estimated as relative frequencies of data at $\xi = 0$ and $\xi = \overline{e}$, respectively, while the continuous part of the *pdf* is expressed as

$$f^C(\xi) = \kappa f^R(\xi), \tag{2.16}$$

where $\kappa = 1 - \kappa_0 - \kappa_1$ and $f^R(\xi)$ is a *pdf* estimated from the subset of data in the open interval $(0, \overline{e})$. Concerning the choice of the structure for the *pdf* $f^R(\xi)$, a mixture of a beta *pdf* and a truncated gamma *pdf* is introduced next.

2.3.1 Beta-Gamma Mixture

The beta distribution is a family of continuous probability distributions with finite support, and is therefore suitable to model random variables assuming values in bounded intervals. In the case of interest, the beta *pdf* over the interval $(0, \overline{e})$ takes the form

$$f_B(\xi) = \begin{cases} \kappa_B \xi^{\alpha-1}(\overline{e} - \xi)^{\beta-1} & \text{if } 0 < \xi < \overline{e} \\ 0 & \text{otherwise,} \end{cases} \tag{2.17}$$

with $\alpha, \beta > 0$ the parameters that control the shape of the distribution, and κ_B a normalizing constant. The gamma distribution is another widely used family of continuous probability distributions. Since the support of the gamma distribution is on $(0, +\infty)$, a truncated version of the gamma *pdf* is considered:

$$f_G(\xi) = \begin{cases} \kappa_G \xi^{k-1} e^{-\xi/\theta} & \text{if } 0 < \xi < \overline{e} \\ 0 & \text{otherwise,} \end{cases} \tag{2.18}$$

with $k > 0$ a shape parameter, $\theta > 0$ a scale parameter and κ_G a normalizing constant. In this chapter, the beta-gamma (BGM) *pdf* is defined as the mixture of a beta *pdf* and a truncated gamma *pdf* with mixing parameter ζ such that $0 \leq \zeta \leq 1$:

$$f_{\mathrm{BGM}}(\xi) = \zeta f_{\mathrm{B}}(\xi) + (1 - \zeta) f_{\mathrm{G}}(\xi). \tag{2.19}$$

Let the BGM model be the *cdf* obtained by integrating the discrete-continuous *pdf* (2.14)–(2.16), where $f^R(\xi)$ in (2.16) takes the form of a BGM *pdf*. Maximum likelihood parameter estimation, performed over several data sets of hourly generated energy e_t, has shown that the BGM model is able to fit the empirical distributions of e_t with higher accuracy than using the unmixed beta and truncated gamma *pdf*s in place of $f^R(\xi)$. For illustration purposes, Fig. 2.2 shows the very good fit of an empirical wind energy *cdf* obtained using the BGM model. Complete results for the 800 kW wind turbine of this example are presented in Sect. 2.5.2.

2.4 Exploiting Wind Speed Forecasts

In Sect. 2.2 contracts were determined assuming only the knowledge of the prior wind energy statistics. In this section, a procedure to exploit day-ahead forecasts $\hat{v}_t$ of hourly average wind speed (e.g., provided by a meteorological service) in the bidding strategy is investigated. The most intuitive approach would be to offer the predicted wind energy profile computed using wind speed forecasts. However, offering wind energy forecasts may lead to unsatisfactory performance for the WPP. With this motivation, an alternative approach which combines classification methods based on machine learning techniques and the optimal bidding strategy of Sect. 2.2 is proposed. Wind speed forecasts are exploited to classify the day of the bidding into one of several predetermined classes. Then, bids are represented by the optimal contracts computed as in Proposition 1 for the selected class.

2.4.1 Offering Wind Energy Forecasts

Wind energy forecasting is a challenging problem which has recently attracted increasing attention from researchers. The interested reader is referred to the survey paper [19] for a review and categorization of different approaches. In many cases, the focus is on wind speed forecasts, which are then converted to power forecasts through the power curve of the wind turbine. This is the approach taken as a reference for the comparisons of different bidding strategies in Sect. 2.5. By plotting the hourly energy ξ generated by a wind turbine versus the hourly average wind speed v, it can be observed that the plotted points can be well approximated by a sigmoid function saturated below at 0 and above at $\overline{e}$. There exist several expressions for sigmoid functions. The one considered in this chapter has the following form:

$$E^S(v) = \omega_2 + (\omega_1 - \omega_2)\left(1 + e^{\frac{v-v_0}{\omega_3}}\right)^{\omega_4}, \qquad (2.20)$$

where $\omega_1 > 0$, $\omega_2 < 0$, $\omega_3 < 0$, $\omega_4 < 0$ and $v_0 > 0$ represent the parameters of the sigmoid function. By using (2.20), the wind energy model can be expressed as:

$$E(v) = \begin{cases} \min(\max(0, E^S(v)), \bar{e}) & \text{if } v \le v_{off} \\ 0 & \text{otherwise,} \end{cases} \qquad (2.21)$$

where v_{off} is a threshold which takes into account high wind speed shutdown of the wind turbine. The parameters $\omega_1, \omega_2, \omega_3, \omega_4, v_0$ and v_{off} are typically estimated from recorded measurements by solving a nonlinear least squares problem. An example is reported in Fig. 2.3, which shows the model (2.21) fitted to the data of a 800 kW wind turbine.

If forecasts of hourly average wind speed $\hat{v}_t$ are available, hourly energy bids can be easily formed by offering the wind energy forecasts computed by substituting $\hat{v}_t$ in (2.21), i.e.,

$$b_t = E(\hat{v}_t), \quad t = 1, \ldots, 24. \qquad (2.22)$$

2.4.2 Day Classification Based on Wind Speed Forecasts

The bidding strategy based on offering wind energy forecasts has a number of drawbacks. First, inaccurate wind speed forecasts may cause unacceptable errors when predicting wind energy through (2.21). Second, and most importantly, the bids (2.22) do not take into account the penalties λ_t^- and λ_t^+. This implies that bidding these forecasts may not be the best offer one can do [8]. In order to mitigate the effects of inaccurate wind speed forecasts and, simultaneously, take explicitly into account the imbalance penalties, in this chapter the idea to combine the optimal bidding strategy described in Sect. 2.2 with a suitable classification strategy based on wind speed forecasts is proposed. Roughly speaking, the proposed approach consists in training a classifier which maps a day (represented by the corresponding wind speed forecasts) to one of several classes associated to different levels of daily generated energy. Then, the bids made for every hour of that day are the optimal contracts computed as in Proposition 1, but using the *conditional* wind energy *cdf* of the corresponding class.

Group the hours of one day into g sets $\mathcal{D}_i = \{a_{i-1} + 1, \ldots, a_i\}$, $i = 1, \ldots, g$, where the integers $a_i \in \mathbb{N}$ are such that

$$0 = a_0 < a_1 < \cdots < a_g = 24.$$

To ease the notation, it is assumed that the sets $\mathcal{D}_i$ have the same cardinality, i.e., they contain the same number h of hours. Denote by $e^{max} = h\bar{e}$ an upper bound on the amount of energy that the wind power plant generates during the hours associated to any set $\mathcal{D}_i$. The energy interval $[0,\ e^{max})$ is then partitioned into ℓ contiguous, non overlapping intervals $[E_{j-1},\ E_j), j = 1,\ \ldots,\ \ell$, such that

$$0 = E_0 < E_1 < \cdots < E_\ell = e^{max}.$$

Now, each day can be classified according to the amount of energy generated during each period $\mathcal{D}_i$. For example, consider the energy generated during "morning" and "evening" hours ($g = 2$, $a_1 = 12$). If only two levels of generated energy ($\ell = 2$) are considered, e.g., "high" ($\uparrow$) and "low" ($\downarrow$) generation, then each day belongs to one of four possible classes $\{(\downarrow, \downarrow), (\downarrow, \uparrow), (\uparrow, \downarrow), (\uparrow, \uparrow)\}$ where the pair (X, Y) denotes the class of days having a X generation level during the first 12 h and a Y generation level during the second 12 h. In general, given g sets of hours and ℓ levels of energy generation, it is possible to define a family of ℓ^g classes $\mathcal{C} = \{\mathcal{C}_{j_1,\ldots,j_g}, j_i = 1, \ldots, \ell,\ i = 1, \ldots, g\}$ such that a day d belongs to the class $\mathcal{C}_{j_1,\ldots,j_g}$, if

$$\sum_{t=a_{i-1}+1}^{a_i} e_t^d \in [E_{j_i-1}, E_{j_i}), \quad i = 1, \ldots, g, \tag{2.23}$$

where e_t^d denotes the wind energy generated during the hour t of day d.

The energy delivered in each period $\mathcal{D}_i$ can be computed only a posteriori. Since the bids must be made in advance, day d is classified a priori on the basis of the corresponding wind speed forecasts $\hat{v}_t^d$. To this aim, an automatic classifier is adopted, which takes as inputs the wind speed forecasts and returns the class the next day will likely belong to. The automatic classifier is trained using a training set created from past data of generated energy and wind speed forecasts. First, each day d of the training set is assigned to the corresponding true class $\mathcal{C}^d \in \mathcal{C}$ on the basis of the actual generated energy. Then, a g-dimensional feature vector representative of the considered day is built from wind speed forecasts. Features are selected as a function of the cube of the wind speed forecasts. Specifically, given the wind speed forecasts $\hat{v}_t^d$ for each hour t of day d, the feature vector $\boldsymbol{f}^d = [f_1^d, \ldots, f_g^d]' \in \mathcal{F} \subseteq \mathbb{R}^g$ is computed, where

$$f_i^d = \sum_{t=a_{i-1}+1}^{a_i} \left(\hat{v}_t^d\right)^3, \quad i = 1, \ldots, g. \tag{2.24}$$

This choice is motivated by the fact that the power that can be extracted from the wind is proportional to the cube of the wind speed. At this point, the training set

$$\mathbf{T} = \left\{\left(\mathcal{C}_{i_1,\ldots,i_g}^d, \boldsymbol{f}^d\right), \ d = 1, \ldots, D_T\right\} \tag{2.25}$$

contains the pairs (class, feature) for each day d, where $\mathcal{C}_{i_1,\ldots,i_g}^d$ denotes the class that day d belongs to, and D_T is the cardinality of the training set. The set $\mathbf{T}$ is used to train a classifier $\Lambda : \mathcal{F} \to \mathcal{C}$ which, given a feature $\boldsymbol{f} \in \mathcal{F}$, returns a class $\Lambda(\boldsymbol{f}) \in \mathcal{C}$. Several approaches can be adopted to identify the function Λ [23]. In this work, the

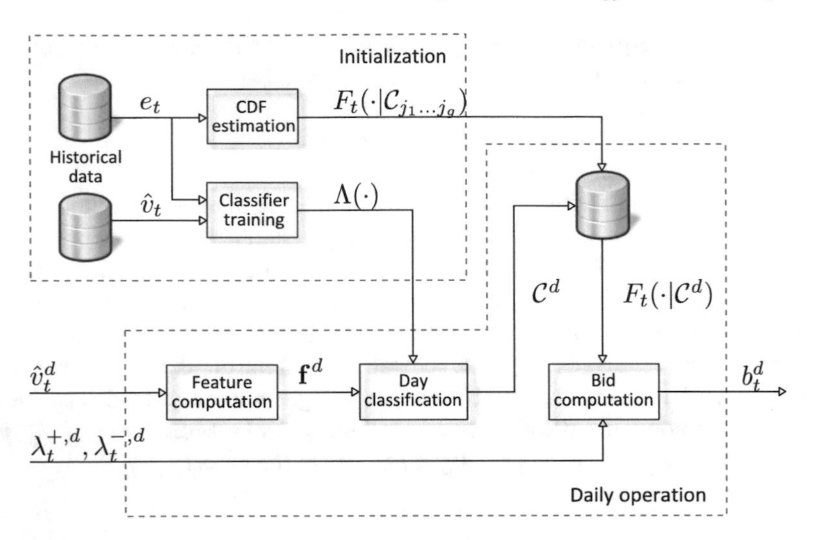

Fig. 2.4 Block scheme of the proposed bidding procedure. It consists of two main parts: "Initialization" and "Daily operation". At initialization, the conditional *cdf*s $F_t(\cdot|\mathcal{C}_{j_1 \ldots j_g})$ and the classifier $\Lambda(\cdot)$ are estimated based on historical data. During daily operation, at day $d-1$ features $\boldsymbol{f}^d$ for day d are first computed from day-ahead wind speed forecasts $\hat{v}_t^d$ according to (2.24). Day d is then associated to the class $\mathcal{C}^d = \Lambda(\boldsymbol{f}^d)$, and the bids b_t^d are computed given the penalties $\lambda_t^{-,d}$ and $\lambda_t^{+,d}$ (or available forecasts), and the conditional *cdf*s $F_t(\cdot|\mathcal{C}^d)$, according to Proposition 1

Multicategory Robust Linear Programming (MRLP) approach described in [24] is adopted.

Having the classifier Λ available, the last step is to determine the optimal bidding strategy for each of the classes $\mathcal{C}_{i_1,\ldots,i_g} \in \mathcal{C}$. This boils down to substituting the *cdf* $F_t(\xi)$ in (2.7) with the conditional *cdf* $F_t(\xi \,|\, \mathcal{C}_{i_1,\ldots,i_g}) = \Pr(e_t \leq \xi \,|\, \mathcal{C}_{i_1,\ldots,i_g})$ for each class $\mathcal{C}_{i_1,\ldots,i_g}$, where $\Pr(\cdot \,|\, \mathcal{C}_{i_1,\ldots,i_g})$ means that the statistics is restricted only to those days belonging to the class $\mathcal{C}_{i_1,\ldots,i_g}$.

The block scheme of the proposed bidding procedure exploiting wind speed forecasts is reported in Fig. 2.4. Note that, to suitably select the classification parameters g, ℓ, a_i and E_j, one has to take into account the overall performance of the bidding procedure in terms of the average daily profit. For instance, too many classes may cause poor estimation of conditional *cdf*s due to small cardinality of data belonging to each class. On the other hand, too few classes may not lead to significant improvements as compared to the bidding strategy of Proposition 1 (corresponding to $g = \ell = 1$).

Remark 4 The bidding strategy presented in Sect. 2.2, as well as the classification method developed in Sect. 2.4 can be extended to different renewable power producers. In fact, the bidding process for PV power producers, dealing with non-stationarity of solar PV energy distribution, has been addressed in [20, 21].

2.5 Simulated Results

In this section, the proposed bidding strategies are validated using real data from a wind turbine installed in Southern Italy. Let OBσ denote the optimal bidding strategy of Proposition 1, where OB stands for optimization-based and σ refers to the use of the tolerance threshold. The basic bidding strategy in [10], which is obtained from Corollary 1, is denoted by OB. The bidding strategy described in Sect. 2.4.1, which uses wind speed forecasts (WF) along with the plant energy curve (PC), is denoted by WF+PC. Finally, the bidding strategy proposed in Sect. 2.4.2, which combines the use of wind speed forecasts for day classification and Proposition 1, is denoted by WF+OBσ.

2.5.1 *Data Set*

A wind turbine of 800 kW nominal power is considered. Hence, $\overline{e} = 800$ kWh. The following data are available:

- generated energy e_t;
- wind speed v_t;
- day-ahead wind speed forecasts $\hat{v}_t$.

The time interval spanned by the data set of generated energy ranges from March 2010 to April 2012. Real and forecast wind speeds are available from September 2011 to April 2012. The data set is split into a training set (from March 2010 to January 2012) and a validation set, containing the remaining data. The energy price λ_t is the clearing price of the Italian day-ahead electricity market, whose time series is public [25]. Penalties λ_t^- and λ_t^+ are randomly simulated under two different scenarios:

- Scenario I: $\lambda_t^-/\lambda_t \sim \mathcal{U}(0.1, 0.3)$, $\lambda_t^+/\lambda_t \sim \mathcal{U}(0.4, 0.6)$,
- Scenario II: $\lambda_t^-/\lambda_t \sim \mathcal{U}(0.4, 0.6)$, $\lambda_t^+/\lambda_t \sim \mathcal{U}(0.8, 1)$,

where $\mathcal{U}(a, b)$ denotes the uniform distribution over the interval $[a, b]$ and the notation $\sim$ means that a random variable follows the specified distribution. Note that larger penalties are applied in Scenario II. Moreover, in both scenarios, $(\lambda_t^-, \lambda_t^+) \in \mathfrak{R}_2$, and therefore the bidding problem consists in the non-trivial case of Proposition 1.

2.5.2 *Estimation of the Wind Power Distributions*

The distributions $F_t(\xi)$ of the generated energy e_t are modeled through the BGM model, whose parameters are estimated using the whole training set as described in Sect. 2.3. In particular, given generation data in the open interval $(0, \overline{e})$ for the hour t of the day, the parameters α, β, k, θ and ζ of the BGM *pdf* (2.17)–(2.19)

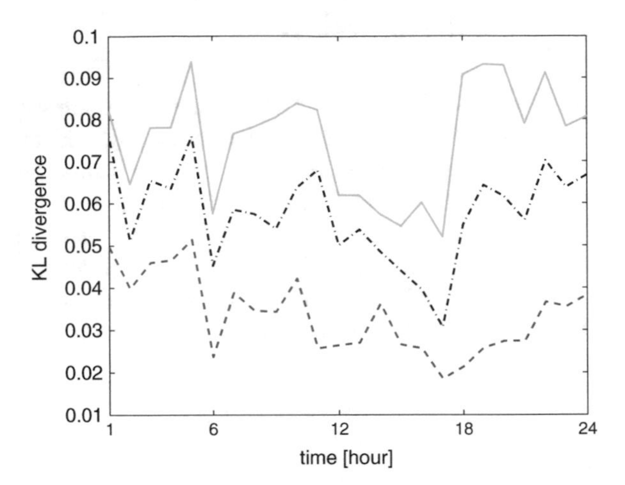

Fig. 2.5 KL divergence of the beta (solid red), truncated gamma (dash-dotted black) and BGM (dashed blue) distributions from the empirical distribution

are estimated using maximum likelihood. For comparison purposes, also a beta *pdf* (2.17) and a truncated gamma *pdf* (2.18) are estimated over the same data using maximum likelihood. Then, the corresponding *cdf*s are obtained by integrating the *pdf*s.

The comparison of the estimated distributions is performed by computing the Kullback-Leibler (KL) divergence of the BGM, beta and truncated gamma distributions from the empirical distribution for all hours of the day. The KL divergence is a well-known, non-symmetric measure of the difference between two probability distributions F and G. For continuous and differentiable distributions, the KL divergence of G from F, where F typically represents the "true" distribution and G is an approximation of F, is defined as

$$D_{KL}(F\|G) = \int_{-\infty}^{+\infty} f(x) \ln \frac{f(x)}{g(x)} \, dx, \tag{2.26}$$

where f and g denote the *pdf*s of F and G, respectively. Values of the KL-divergence from the empirical distribution are plotted in Fig. 2.5. It can be observed that the KL divergence of the BGM distribution is always significantly below the KL divergence of the other two distributions.

2.5.3 Computation of the Bids

For the bidding strategies OB and OBσ, the whole training set is used to estimate the distributions $F_t(\xi)$ of the generated energy e_t. Then, for given penalties $(\lambda_t^-, \lambda_t^+) \in \mathfrak{R}_2$ and tolerance σ, the bids b_t are computed by solving (2.7). For fixed t, the bid b_t is shown in Fig. 2.6 as a function of σ for different values of φ_t in (2.8). It can be

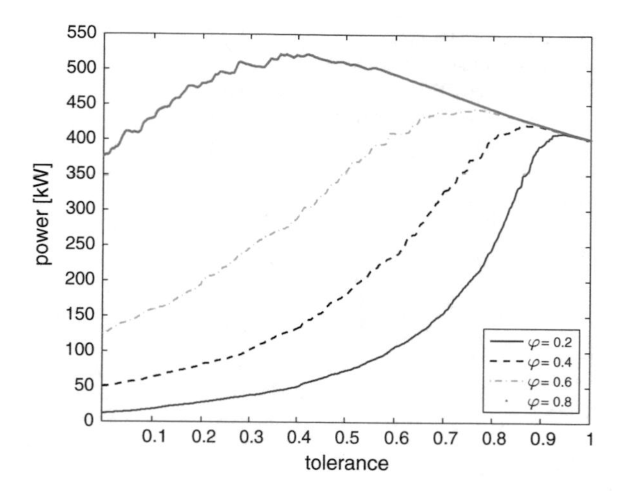

Fig. 2.6 Bid b_t computed by solving (2.7) as a function of the tolerance σ for different values of φ_t and $t = 20$

observed that, for all values of φ_t, b_t tends to $\overline{e}/2$ as σ tends to 1. Indeed, when $\sigma = 1$ and $b_t = \overline{e}/2$, the whole hourly energy interval $[0, \overline{e}]$ is covered by the tolerance band.

Concerning the bidding strategy WF+PC, data points (v_t, e_t) available in the training set are used to estimate the energy curve (2.21) of the wind turbine by solving a nonlinear least squares problem. The energy curve fitted to the data for the considered wind turbine is shown in Fig. 2.3. In the validation phase, the estimated energy curve and the wind speed forecasts $\hat{v}_t$ are used to compute the bids b_t through (2.22).

In the case of the bidding strategy WF+OBσ, classification parameters g, ℓ, a_i and E_j are selected so as to maximize the average daily profit over estimation data (see the end of Sect. 2.4.2). It turns out that a suitable choice is to classify each day on the basis of the energy generated during the first and the second 12 hours. For each 12-hour period, two levels of generation are defined, with the threshold set at 27% of the maximum amount of energy that the wind turbine may generate over the period. According to the notation of Sect. 2.4, this means to set $g = 2$, $a_1 = 12$, $\ell = 2$, $E_1 = 0.27e^{max}$, where $e^{max} = 12\overline{e} = 9.6$ MWh, resulting in a family of four classes $\mathcal{C} = \{\mathcal{C}_{11}, \mathcal{C}_{12}, \mathcal{C}_{21}, \mathcal{C}_{22}\}$. Given these classes, the conditional distributions $F_t(\xi|\mathcal{C}_{ij})$ are estimated as described in Sect. 2.3, using only the generated energy e_t in days belonging to class $\mathcal{C}_{ij}$, $i, j = 1, 2$. Then, for given penalties $(\lambda_t^-, \lambda_t^+) \in \mathfrak{R}_2$ and tolerance σ, the bids b_t are computed for each class solving (2.7), where φ_t is given by (2.8). Differences between the daily bidding profiles thus obtained for the four classes can be appreciated in Fig. 2.7. To complete the bidding strategy WF+OBσ, a classifier Λ is needed, as described in Sect. 2.4.2. To this purpose, for each day d in the training set, the feature vector $\boldsymbol{f}^d = [\, f_1^d \; f_2^d \,]'$ is computed as in (2.24). Pairs (class, feature) are then used to train a MRLP classifier by solving a linear program to minimize the number of misclassifications [24]. The regions in the feature space resulting from the training of the MRLP classifier are shown in Fig. 2.8. It is worth noting that the training only takes few seconds on a standard desktop PC. Consequently, it can be repeated whenever it is deemed necessary (e.g.,

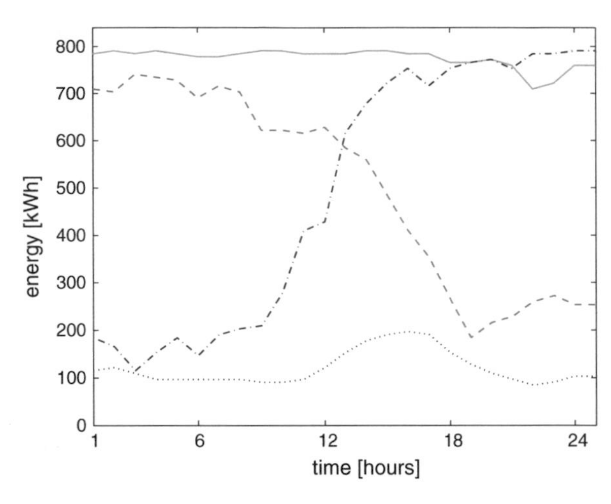

Fig. 2.7 Daily bidding profiles computed using the bidding strategy WF+OBσ with fixed $\varphi_t = 0.71$ for all t and $\sigma = 0$: classes $\mathcal{C}_{11}$ (dotted black), $\mathcal{C}_{12}$ (dot-dashed blue), $\mathcal{C}_{21}$ (dashed red) and $\mathcal{C}_{22}$ (solid green)

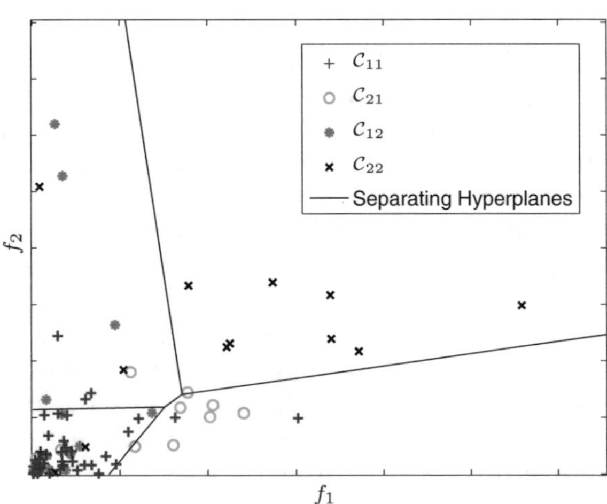

Fig. 2.8 Classification of the feature vectors and regions resulting from the training of the MRLP classifier

for incorporating additional information collected over the days) without introducing any delay in the bidding process.

2.5.4 *Comparison of the Bidding Strategies*

In this section, the considered bidding strategies are compared under the assumption that the penalties λ_t^- and λ_t^+ are known beforehand, so that their true values can be used to compute the bids according to OB, OBσ and WF+OBσ. This is done in order to highlight the potential benefits of using wind speed forecasts in conjunction with the optimal bidding strategy (2.6). Performance degradation due to forecasting errors on λ_t^- and λ_t^+ is analyzed in the next subsection. For all bidding strategies,

Table 2.1 Average daily profit (€) in Scenario I with known penalties

Strategy\\σ	0	0.05	0.1	0.15	0.2
OB	365	372	380	387	395
OBσ	365	376	389	402	415
WF+PC	375	379	384	387	391
WF+OBσ	432	440	447	454	460

Table 2.2 Average daily profit (€) in Scenario II with known penalties

Strategy\\σ	0	0.05	0.1	0.15	0.2
OB	172	183	194	204	215
OBσ	172	191	211	232	254
WF+PC	239	248	256	263	269
WF+OBσ	313	329	343	356	368

given the bids b_t, the generated energy e_t, the energy prices λ_t and the penalties λ_t^- and λ_t^+, the net hourly profits (2.1) are evaluated, aggregated at daily level and then averaged over all days in the validation data set. Average daily profits for different values of the tolerance σ are reported in Table 2.1 for Scenario I and Table 2.2 for Scenario II.

As expected, in both of the considered scenarios, OBσ performs better than OB when $\sigma > 0$ (for $\sigma = 0$ the two bidding strategies coincide). This confirms the importance of adapting the bidding strategy to the tolerance σ. The advantage of using OBσ is stronger in Scenario II, where penalties are higher. Indeed, for $\sigma = 0.2$, OBσ improves 5.06% with respect to OB in Scenario I and 18.14% in Scenario II.

Comparison of the performance of OBσ and WF+OBσ, which are both tailored to the tolerance σ, shows that WF+OBσ outperforms OBσ in both scenarios and for all considered values of σ. The improvement of WF+OBσ over OBσ ranges from 10.84% for $\sigma = 0.2$ to 18.36% for $\sigma = 0$ in Scenario I, and from 44.89% for $\sigma = 0.2$ to 82.00% for $\sigma = 0$ in Scenario II. Note that the advantage of using WF+OBσ is stronger in Scenario II, where penalties are higher. Moreover, in both scenarios, the improvement of WF+OBσ over OBσ is bigger as the threshold σ is decreased. The presented results are obtained with a percentage of correct classification around 70% provided by the MRLP classifier. If one could apply an ideal classifier with 100% correct classification, the improvement of WF+OBσ over OBσ would range from 19.28% for $\sigma = 0.2$ to 28.49% for $\sigma = 0$ in Scenario I, and from 72.44% for $\sigma = 0.2$ to 123.26% for $\sigma = 0$ in Scenario II. This shows the potential of using wind speed forecasts for increasing the WPP expected profit, still leaving room for improvements with respect to the presented results. For instance, one might consider classifiers with nonlinear separating hypersurfaces which could guarantee a higher percentage of correct classification than the considered MRLP classifier.

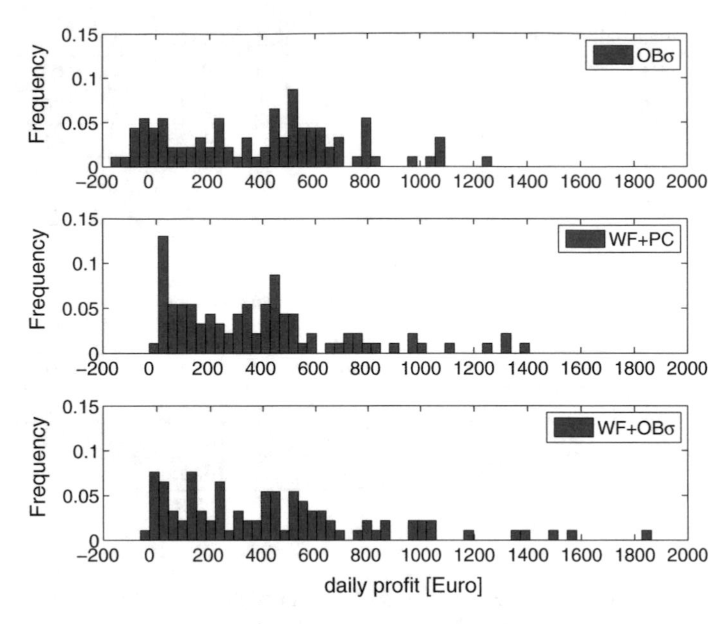

Fig. 2.9 Histograms of the daily profits of the bidding strategies OBσ (top), WF+PC (middle) and WF+OBσ (bottom) for Scenario I and threshold $\sigma = 0.1$

Though WF+PC and WF+OBσ exploit the same wind speed forecasts to shape the energy bids, WF+PC performs poorly with respect to WF+OBσ in both scenarios and for all considered values of σ. The improvement of WF+OBσ over WF+PC ranges from 15.20% for $\sigma = 0$ to 17.65% for $\sigma = 0.2$ in Scenario I, and from 30.96% for $\sigma = 0$ to 36.80% for $\sigma = 0.2$ in Scenario II. These results are obtained by using commercial wind speed forecasts (rather coarse and inaccurate) averaged at a regional level. These inaccurate forecasts penalize directly WF+PC, which requires to evaluate the plant energy curve at each given wind speed. On the other hand, the classification strategy proposed in Sect. 2.4.2 is less sensitive to forecast uncertainties. Indeed, both the "integration" over time made when computing the features (2.24) and the fact that the classifier extracts discretized information from the feature vectors, contribute to filter out the uncertainty and "robustify" the technique.

For comparison purposes, the average daily profit of the ideal strategy denoted by R, consisting in an oracle which offers the exact amount of energy generated the next day, is also considered. Since $b_t = e_t$, the producer never incurs penalties, and therefore, being $(\lambda_t^-, \lambda_t^+) \in \mathfrak{R}_2$, the profit is maximal. In this way, the performance of the compared bidding strategies can be evaluated with respect to the maximum achievable. With the bidding strategy R, the average daily profit would be 542 €. This implies that applying WF+OBσ allows one to fill between 35.43 and 40.00% of the gap between OBσ and R in both scenarios and for all considered values of σ.

An example of the distributions of the daily profit, whose mean values are reported in Tables 2.1 and 2.2, is presented in Fig. 2.9 for the bidding strategies OBσ, WF+PC

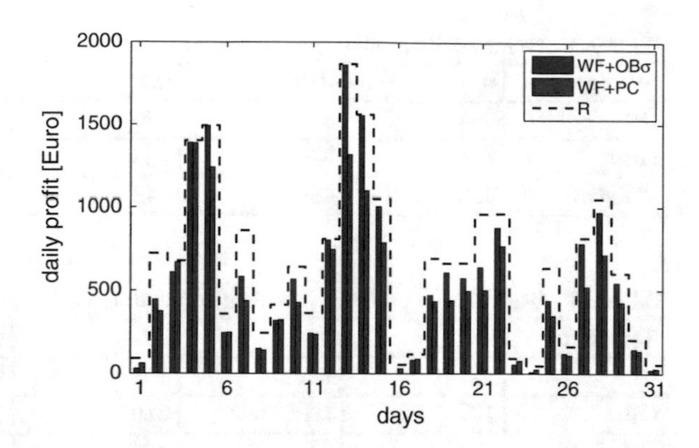

Fig. 2.10 Bar plots of the daily profits of the bidding strategies WF+PC and WF+OBσ for Scenario I and threshold $\sigma = 0.1$. The maximum daily profit obtained by applying the ideal bidding strategy R is also reported (dashed)

and WF+OBσ under Scenario I and threshold $\sigma = 0.1$. The histogram of OBσ ranges from -135 to 1257 €, with mean value 389 €. OBσ provides a negative daily profit in 15% of the days in the validation period. The histogram of WF+PC ranges from around 0 to 1388 €, with mean value 384 €. Though OBσ and WF+PC have comparable average performance, WF+PC would be preferable in this case because it never provides a negative daily profit. Finally, the histogram of WF+OBσ ranges from -36 to 1860 €, with mean value 447 €. WF+OBσ provides a negative daily profit in 9% of the days. However, the difference between WF+PC and WF+OBσ is negligible in all these days, being always less than 40 €. In 15% of the days, WF+OBσ provides a positive daily profit which is still below the daily profit provided by WF+PC. The difference between WF+PC and WF+OBσ never exceeds 61 € in these cases. In all the remaining 76% of the days, WF+OBσ performs better than WF+PC with an average improvement of 90 € per day. Direct comparison of the daily profits provided by WF+PC and WF+OBσ over the first 31 days of the validation period is done in Fig. 2.10, which also shows the daily profits provided by the ideal strategy R. It can be noticed that WF+OBσ is in many days very close to the maximum daily profit achievable.

2.5.5 Sensitivity Analysis Under Unknown Penalties

In this section, the sensitivity of the bidding strategies OB, OBσ and WF+OBσ to uncertain penalties λ_t^- and λ_t^+ is evaluated. The penalties of Scenario I and Scenario II are considered as the true penalties (known only a posteriori), and used to evaluate the hourly profit (2.1). On the other hand, the bids are computed according to OB, OBσ and WF+OBσ by replacing the true penalties with their expected values. Given the distributions of λ_t^-/λ_t and λ_t^+/λ_t assumed in Sect. 2.5.1, this results into constant values $\varphi_t = 0.71$ in Scenario I and $\varphi_t = 0.64$ in Scenario II, for all t. It is stressed that this choice may lead to bids that are very different from those computed using

Table 2.3 Average daily profit (€) in Scenario I with unknown penalties

Strategy\\σ	0	0.05	0.1	0.15	0.2
OB	364	371	379	386	393
OBσ	364	376	388	400	414
WF+OBσ	428	436	443	450	457

Table 2.4 Average daily profit(€) in Scenario II with unknown penalties

Strategy\\σ	0	0.05	0.1	0.15	0.2
OB	171	182	193	203	213
OBσ	171	190	210	231	254
WF+OBσ	311	328	342	355	367

the true penalties (e.g., φ_t ranges from 0.57 to 0.86 in Scenario I, if the true penalties are used).

Average daily profits for different values of the tolerance σ are reported in Table 2.3 for Scenario I and Table 2.4 for Scenario II in the case of unknown penalties when the bids are computed. Bidding strategy WF+PC does not appear in the two tables since it is independent of the values of the penalties. Comparing Table 2.1 with Table 2.3 and Table 2.2 with Table 2.4, it can be seen that the average daily profits obtained in the cases of known and unknown penalties differ by a small amount (less than 4 €) for all considered bidding strategies and values of σ. This indicates a low sensitivity of OB, OBσ and WF+OBσ to uncertain penalties.

2.6 Conclusions

In this chapter, the problem of computing the optimal day-ahead bidding profile for a WPP is addressed in a framework where penalties are applied only outside tolerance bands around the nominal contracts. The optimization of the day-ahead bidding profile is embedded in a day classification approach which exploits wind speed forecasts provided by a meteorological service to classify the next day. In this way, a suitable bidding profile can be offered according to the predicted class. A numerical comparison of different bidding strategies is performed on real data from an Italian wind plant, showing that exploiting wind speed forecasts through classification allows one to improve consistently the profit of the WPP, with respect both to the case where no classification is adopted and to the case in which the day-ahead bidding profile is computed according to the wind speed forecasts and the plant energy curve.

References

1. Giannitrapani A, Paoletti S, Vicino A, Zarrilli D (2013) Wind power bidding in a soft penalty market. In: Proceedings of IEEE conference on decision and control, pp 1013–1018
2. Giannitrapani A, Paoletti S, Vicino A, Zarrilli D (2016) Bidding wind energy exploiting wind speed forecasts. IEEE Trans Power Syst 31(4):2647–2656
3. Denny E, O'Malley M (2006) Wind generation, power system operation, and emissions reduction. IEEE Trans Power Syst 21(1):341–347
4. Barth R, Weber C, Swider DJ (2008) Distribution of costs induced by the integration of RES-E power. Energy Policy 36(8):3107–3115
5. Usaola J, Ravelo O, González G, Soto F, Dávila MC, Díaz-Guerra B (2004) Benefits for wind energy in electricity markets from using short term wind power prediction tools; a simulation study. Wind Eng 28(1):119–127
6. Holttinen H (2005) Optimal electricity market for wind power. Energy Policy 33(16):2052–2063
7. Bremnes JB (2004) Probabilistic wind power forecasts using local quantile regression. Wind Energy 7(1):47–54
8. Pinson P, Chevallier C, Kariniotakis GN (2007) Trading wind generation from short-term probabilistic forecasts of wind power. IEEE Trans Power Syst 22(3):1148–1156
9. Dent CJ, Bialek JW, Hobbs BF (2011) Opportunity cost bidding by wind generators in forward markets: analytical results. IEEE Trans Power Syst 26(3):1600–1608
10. Bitar EY, Rajagopal R, Khargonekar PP, Poolla K, Varaiya P (2012) Bringing wind energy to market. IEEE Trans Power Syst 27(3):1225–1235
11. Matevosyan J, Söder L (2006) Minimization of imbalance cost trading wind power on the short-term power market. IEEE Trans Power Syst 21(3):1396–1404
12. Matevosyan J, Söder L (2010) Short-term trading for a wind power producer. IEEE Trans Power Syst 25(1):554–564
13. Jónsson T, Pinson P, Madsen H (2010) On the market impact of wind energy forecasts. Energy Econ 32(2):313–320
14. Botterud A, Wang J, Bessa RJ, Keko H, Miranda V (2010) Risk management and optimal bidding for a wind power producer. In: Proceedings of IEEE PES general meeting, pp 1–8
15. Botterud A, Wang J, Bessa RJ, Keko H, Miranda V (2013) Optimal bidding strategies for wind power producers with meteorological forecasts. In: Proceedings of SIAM conference on control and its applications, pp 13–20
16. Morales JM, Conejo AJ, Pérez-Ruiz J (2009) Economic valuation of reserves in power systems with high penetration of wind power. IEEE Trans Power Syst 24(2):900–910
17. Fabbri A, Gómez San Román T, Abbad JR, Quezada VHM (2005) Assessment of the cost associated with wind generation prediction errors in a liberalized electricity market. IEEE Trans Power Syst 20(3):1440–1446
18. Sideratos G, Hatziargyriou ND (2007) An advanced statistical method for wind power forecasting. IEEE Trans Power Syst 22(1):258–265
19. Ma L, Luan S, Jiang C, Liu H, Zhang Y (2009) A review on the forecasting of wind speed and generated power. Renew Sustain Energy Rev 13(4):915–920
20. Giannitrapani A, Paoletti S, Vicino A, Zarrilli D (2013) Exploiting weather forecasts for sizing photovoltaic energy bids. In: Proceedings of IEEE PES innovative smart grid technologies European conference, pp 1–5
21. Giannitrapani A, Paoletti S, Vicino A, Zarrilli D (2014) Bidding strategies for renewable energy generation with non stationary statistics. In: Proceedings of IFAC world congress, pp 10,784–10,789
22. Bitar EY, Rajagopal R, Khargonekar PP, Poolla K (2011) The role of co-located storage for wind power producers in conventional electricity markets. In: Proceedings of American control conference, pp 3886–3891

23. Boucheron S, Bousquet O, Lugosi G (2005) Theory of classification: a survey of some recent advances. ESAIM Probab Stat 9:323–375
24. Bennett K, Mangasarian O (1994) Multicategory discrimination via linear programming. Optim Methods Softw 3:27–39
25. Gestore dei Mercati Energetici, Web site of the Italian Electricity Market Managing Company. http://www.mercatoelettrico.org

Chapter 3
Demand Response Management in Smart Buildings

In this chapter an optimization approach based on Model Predictive Control (MPC) for allowing the temperature control system of large buildings to participate in a DR program is proposed. The material of this chapter is mainly based on [1, 2].

3.1 Introduction

Building energy consumption, both commercial and residential, represents almost 40% of the global energy produced worldwide. About 50% of this huge amount of energy is consumed for heating, ventilation and air conditioning (HVAC). HVAC plants, especially the oldest ones, are operated through simple rule-based strategies, which actuate the system in feedforward at a centralized level, while local thermal control is made through standard thermostatic devices. The need to improve these out-of-date techniques has stimulated research for several years, with the aim of reducing consumption and improving comfort through the design of more appropriate and sophisticated feedback control laws exploiting real-time information from the several components of the buildings. Most of the proposed approaches are based on MPC, because of its attractive features, ranging from the possibility of handling constraints on numerous variables involved to optimizing economical objectives in a time-varying context (see e.g., [3–7] and references therein). The recent paper [8] provides a comprehensive framework based on MPC and co-simulation for real time control of the energy management system of a building.

An important issue which is gaining much attention in the recent literature on electricity systems and building thermal control concerns DR. The primary aim of DR is to overcome the "traditional" inflexibility of electrical demand and, among other benefits, to maximize deployment of RES. A comprehensive view on technical methodologies and architectures, commercial arrangements, and socio-economic factors that could facilitate the uptake of DR has been addressed in [9]. In this context,

D. Zarrilli, *Integration of Low Carbon Technologies in Smart Grids*,
Springer Theses, https://doi.org/10.1007/978-3-319-98358-5_3

building operators can be considered as excellent candidates for demand flexibility, as they might find it convenient to schedule certain tasks in order to obtain a reward. In particular, the possibility of shifting HVAC electric loads according to a smart strategy is crucial to participation in DR services. In [10], the impact of buildings in DR programs on the electricity market is modeled through an agent-based simulation platform, and it is shown how different levels of DR penetration affect the market prices. An approach for allocating the requested energy among heterogeneous devices in buildings depending of DR requests and electricity price has been reported in [11].

In the present chapter, an optimization approach based on MPC for allowing the temperature control system of large buildings to participate in a DR program is proposed. The idea of exploiting MPC for supervisory control of building energy management systems traces back to the late eighties [12], even if the intrinsic computational burden of the approach prevented realistic applications until a few years ago. Participation of buildings in DR programs has been recently addressed in [13], where a pricing policy has been proposed for offering real time regulation services, and in [14], where an optimization framework based on genetic algorithms is provided for dealing with a DR case study for the heating of an office building in Canada. In order to rapidly reply to DR requests, a fast chiller power demand response control strategy for commercial buildings is introduced in [15], with the aim of maintaining internal thermal comfort by regulating the chilled water flow distribution under the condition of insufficient cooling supply. A controller for HVAC systems able to curtail peak load while maintaining reasonable thermal comfort has been introduced in [16, 17], where the set-point temperature of a building is changed whenever the retail price is higher than customers preset price.

The novelty introduced in this chapter with respect to the work mentioned above is the integration of price-volume signals provided by an aggregator into the temperature regulation system, and the development of a cost-optimal control strategy involving low computational complexity for DR-enabled large-scale buildings. On the basis of external price-volume signals, the optimizer analyses whether it is convenient for the building to honour the corresponding DR requests. The objective is the minimization of the energy bill. Since the complexity of the overall optimization problem is intractable even for buildings of modest dimension, a suitable heuristic search strategy based on problem decoupling is devised in order to make the computational burden acceptable without significant loss of accuracy. The approach is independent of the particular heating technology adopted and it can be easily generalized to cooling management as well.

The proposed technique is validated using EnergyPlus [18] as a realistic physical modeling simulator. The MPC optimal control problem is solved on the basis of an identified linear model of the building. The control law is tested on the linear model and on the physical model simulator for comparison purposes. In this sense, the present contribution is in the spirit of [19–21], where a different objective is considered. It is shown that a decoupling approach which decomposes regulation of the different zones of the building into independent problems provides reliable results in the absence of DR participation. Since the presence of price-volume signals makes the computational burden of the optimization grow exponentially with the number of zones, a decoupled heuristic relaxation of the problem is devised. The obtained results are compared to the optimal solution on a three-zone building equipped with underfloor electric heaters. A test case involving a large-scale building equipped with a heat pump heating system is also worked out.

The chapter is organized as follows. In Sects. 3.2 and 3.3 the building system model and the DR model, are respectively presented. In Sect. 3.4 the heating operation problem is formulated and the proposed control algorithm is described. Section 3.5 reports the experimental results obtained on a small-scale and a large-scale test case, together with a discussion of the results obtained. Finally, conclusions are drawn in Sect. 3.6.

3.2 Building System Model

Consider a building with a centralized heating system (e.g., a government building) composed of z zones $Z_1, \ldots, Z_z$ equipped with electrical heating devices, e.g., electrical radiant floors or heat pumps. Assume that each zone is equipped with a temperature sensor connected to the centralized controller and that each heater can be independently switched on or off by the control unit. Assume that the control system operates in discrete-time with sampling period Δt. Let $\mathcal{T} = \{0, 1, \ldots\}$ be the set of discrete time indices and $t \in \mathcal{T}$ the generic time index. Moreover, define:

- $u_i(t) \in \mathbb{B}$: heater status $\{0 = \text{ inactive}, 1 = \text{ active}\}$ at time t for Z_i,
- w_i: heater energy consumption [kWh per sampling period] for Z_i,
- $T_i^{in}(t)$: indoor temperature [°C] at time t, measured by the sensor for Z_i,
- $\varsigma_i(t) = [\underline{T}_i(t), \overline{T}_i(t)]$: thermal comfort range for Z_i at time t,
- $\mathbf{u}(t) = [u_1(t) \ldots u_z(t)]' \in \mathbb{B}^z$,
- $\mathbf{T}^{in}(t) = [T_1^{in}(t) \ldots T_z^{in}(t)]' \in \mathbb{R}^z$,
- $\lambda(t)$: electricity price at time t, or forecast thereof.

Other than on the heater statuses $\mathbf{u}(t)$, indoor temperatures $\mathbf{T}^{in}(t)$ may depend on exogenous variables like outdoor temperature, solar radiation, indoor lights and appliances, human occupancy, etc. For a given building, available measurements or forecasts of some or all of these variables are collected in a vector $\varepsilon(t) = [\varepsilon_1(t) \ldots \varepsilon_z(t)]'$. Hence, the temperature dynamics can be modeled in regression form as

$$\mathbf{T}^{in}(t+1) = \mathbf{F}(\boldsymbol{\Phi}(t)), \tag{3.1}$$

where the regression matrix $\boldsymbol{\Phi}(t)$ is given by

$$\boldsymbol{\Phi}(t) = [\mathbf{T}^{in}(t) \ \ldots \ \mathbf{T}^{in}(t - n_{\mathbf{T}^{in}}) \ \mathbf{u}(t) \ \ldots \ \mathbf{u}(t - n_{\mathbf{u}}) \ \varepsilon(t) \ \ldots \ \varepsilon(t - n_{\varepsilon})]', \tag{3.2}$$

being $n_{\mathbf{T}^{in}}, n_{\mathbf{u}}, n_{\varepsilon}$ suitable nonnegative integers that define the model orders, and being $\mathbf{F}(\cdot)$ some (possibly nonlinear and time-varying) function. Given the definitions above, thermal comfort at time t is guaranteed whenever $\mathbf{T}^{in}(t) \in \varsigma(t)$ where

$$\varsigma(t) = \left\{\mathbf{T}^{in}(t) \ : \ T_i^{in}(t) \in \varsigma_i(t) \ \ \forall i = 1, \ldots, z\right\}, \tag{3.3}$$

while the overall building consumption within time step t can be computed as

$$r(t) = \sum_{i=1}^{z} w_i u_i(t). \tag{3.4}$$

Consider a generic time horizon $\mathcal{I}(t, H) = [t, t + H) \subseteq \mathcal{T}$, and let $R(t, H)$ denote the total consumption within $\mathcal{I}(t, H)$, i.e.,

$$R(t, H) = \sum_{\tau=t}^{t+H-1} r(\tau). \tag{3.5}$$

The total expected cost of energy in the interval $\mathcal{I}(t, H)$ is therefore given by

$$C(t, H) = \sum_{\tau=t}^{t+H-1} \lambda(\tau) r(\tau). \tag{3.6}$$

3.3 DR Model

The concept of DR has been introduced several years ago in the literature on smart energy grids [22–24]. Recently, a complete commercial and technical architecture has been developed in the European project ADDRESS [25–27]. Computational methods for technical validation of demand response products have been proposed in [28]. The relevance of this concept to the design of efficient energy management systems is testified by several recent papers, like e.g, [29, 30]. The key idea is that the end users play an active role in the electricity system by adjusting their consumption patterns according to dynamic energy pricing policies enforced by the players involved in energy markets. DR participation does not take place on an individual basis, but rather via the aggregation of a community of individual consumers, possibly represented by an intermediary subject, the *aggregator*. The aggregator's main objective is to provide value by employing the *flexibility* of the consumption profile of individual consumers. Its basic role is to collect certain amounts of energy over specified time intervals, e.g., the energy saved by consumers accepting the aggregator's offers. This energy can be used for several purposes. For instance, the distribution system operator (DSO) may ask an aggregator to enforce energy reduction in a given load area over a given time interval if an overload is foreseen in that area, in order to counteract possible network unbalancing. Efficient grid management, in turn, contributes to overall reduction of carbon dioxide production [31]. A further reason for the aggregator to collect energy is that options related to reprofiling of the load curve in specific load areas of the distribution system, can be sold on the market. An aggregator has a pool of subscribers (end users), and is able to send them *price-volume signals* in order to affect their consumption pattern. These signals are typically sent once or twice a day and specify a monetary reward (price) if power consumption, during certain hours of the day, is below or above specified thresholds (volume) [27].

For the purpose of this work, a standard model of DR program is employed. A DR program is a sequence of DR requests $\mathcal{R}_j$, each involving a time horizon $\mathcal{I}(\mu_j, h_j)$, a total energy bound $\overline{r}_j$, and a monetary reward π_j. A single request $\mathcal{R}_j$ is said to be *fulfilled* if the total building consumption within $\mathcal{I}(\mu_j, h_j)$, i.e., $R(\mu_j, h_j)$, is no higher/lower than the prescribed threshold $\overline{r}_j$, and in this case a monetary reward π_j is granted to the building operator. In the following, only the case in which the aggregator coordinates the consumers' demand profile by exploiting their willingness to reduce the consumption is addressed. This is due to the nature of the considered problem, where it seems less likely that heating thermal loads are controlled with the aim of increase electricity consumption.

Definition 1 A DR program $\mathcal{P}$ is a sequence of DR requests $\mathcal{R}_j$, $j = 1, 2, \ldots$, where $\mathcal{R}_j$ is the set

$$\mathcal{R}_j = \left\{\mathcal{I}(\mu_j, h_j), \overline{r}_j, \pi_j\right\}, \tag{3.7}$$

being

$$\mathcal{I}(\mu_j, h_j) \subseteq \mathcal{T}, \;\; \mathcal{I}(\mu_{j_1}, h_{j_1}) \cap \mathcal{I}(\mu_{j_2}, h_{j_2}) = \emptyset, \;\; \forall j_1 \neq j_2. \tag{3.8}$$

The request $\mathcal{R}_j$ is fulfilled if and only if

$$R(\mu_j, h_j) \leq \overline{r}_j. \tag{3.9}$$

It is worth stressing that the same concepts and algorithms developed hereafter can be extended to the situation opposite to (3.9), i.e., $R(\mu_j, h_j) \geq \overline{r}_j$. However, further research would have to be carried out for the case in which DR requests of both types are included in the same horizon.
For any given time horizon $\mathcal{I}(t, H)$, let

$$\mathcal{P}(t, H) = \left\{\mathcal{R}_j : \mathcal{I}(\mu_j, h_j) \subseteq \mathcal{I}(t, H)\right\}, \tag{3.10}$$

be the set of DR requests that occur within the time horizon. Moreover, it is assumed that the length H is dynamically adapted in a way such that an integer number of DR requests falls inside $\mathcal{I}(t, H)$. Let $\mathcal{J}(t, H)$ be the set of indices identifying such DR requests, i.e.,

$$\mathcal{J}(t, H) = \left\{j : \mathcal{R}_j \in \mathcal{P}(t, H)\right\}. \tag{3.11}$$

For each request $\mathcal{R}_j$, introduce a binary variable $u_j^{DR} \in \mathbb{B}$ defined as

$$u_j^{DR} = \begin{cases} 1 \text{ if } \mathcal{R}_j \text{ is fulfilled} \\ 0 \text{ otherwise.} \end{cases} \tag{3.12}$$

The overall expected cost of operation of the building heating system within the time horizon $\mathcal{I}(t, H)$ under the DR program $\mathcal{P}$, is therefore given by

$$C^{\mathcal{P}}(t, H) = C(t, H) - \sum_{j \in \mathcal{J}(t,H)} u_j^{DR} \pi_j, \tag{3.13}$$

i.e., the expected cost of energy minus the total reward for the fulfilled DR requests.

3.4 Optimal Heating Operation Problem

The goal of this section is to devise a control algorithm for the thermal heating system of each zone in order to minimize the building electricity bill under a DR program $\mathcal{P}$, while preserving comfort constraints.

Consider a time horizon $\mathcal{I}(t, H)$ and collect heater activation status variables for all zones within $\mathcal{I}(t, H)$ in the following $H \times z$ binary matrix:

$$\mathbf{U}(t, H) = \begin{bmatrix} \mathbf{u}(t)' \\ \vdots \\ \mathbf{u}(t + H - 1)' \end{bmatrix} \in \mathbb{B}^{H \times z}. \tag{3.14}$$

Assuming $\varepsilon(t)$ (or a forecast thereof) is available, the above problem can be formulated as a mixed-integer program as follows.

Problem 1 Optimal heating control under DR program $\mathcal{P}$.

$$\begin{cases} \mathbf{U}^*(t, H) = \arg \min\limits_{\substack{\mathbf{U}(t,H) \\ u_j^{DR} : j \in \mathcal{J}(t,H)}} C(t, H) - \sum\limits_{j \in \mathcal{J}(t,H)} u_j^{DR} \pi_j \\ \text{s.t.:} \\ R(\mu_j, h_j) \leq u_j^{DR} \overline{r}_j + (1 - u_j^{DR}) h_j \sum_{i=1}^{z} w_i, \quad \forall j \in \mathcal{J}(t, H) & \text{(a)} \\ u_j^{DR} \in \mathbb{B}, \quad \forall j \in \mathcal{J}(t, H) & \text{(b)} \\ \mathbf{T}^{in}(\tau + 1) = \mathbf{F}(\mathbf{\Phi}(\tau)) & \text{(c)} \\ \mathbf{T}^{in}(\tau) \in \varsigma(\tau), \quad \forall \tau \in \mathcal{I}(t, H) & \text{(d)} \\ \mathbf{U}(t, H) \in \mathbb{B}^{H \times z} & \text{(e)} \end{cases} \tag{3.15}$$

In Problems 1, (3.15a) and (3.15b) represent the DR constraints, i.e., $R(\mu_j, h_j) \leq \overline{r}_j$ for each fulfilled $\mathcal{R}_j$. In particular, when $u_j^{DR} = 0$, the building consumption $R(\mu_j, h_j)$ is unconstrained since the term $h_j \sum_{i=1}^{z} w_i$ represents the maximum electricity consumption over h_j. In the opposite situation, in which $u_j^{DR} = 1$, the left-hand side of (3.15a) disappears and the overall consumption is forced to be less than $\overline{r}_j$. Constraints (3.15c) represent the temperature dynamics, (3.15d) are the comfort constraints, while (3.15e) forces the on/off heater behavior. Note that if the map $\mathbf{F}(\cdot)$ is linear, then Problem 1 is a mixed-integer linear program (MILP).

3.4.1 Sub-optimal Control Algorithm

It is known that temperature dynamics in a given zone depends on heater status, outdoor temperature, solar radiation, internal lights and appliances, occupancy, and temperature of neighboring zones. A complex model which takes into account all the above mentioned aspects is not conceivable for Problem 1 due to the unacceptable computational burden. In particular, if the temperature variables $T_i^{in}(t)$ and the binary decision variables $u_i(t)$ and u_j^{DR} are fully coupled via the constraints in (3.15), then the computational complexity scales exponentially with z, thus making the approach totally unfeasible except for very small-scale problems. In order to overcome this limitation, sub-optimal solutions will be derived by suitably decoupling Problem 1 into z smaller problems. To this purpose, the first step is to obtain a decoupled linear regression building model. Therefore, the following assumption, which boils down to neglecting thermal flow between zones, is enforced.

Assumption 1 The temperature dynamics of each zone Z_i is given by

$$T_i^{in}(t+1) = \mathbf{\Phi}_i'(t)\Theta_i, \quad i = 1, \ldots, z, \tag{3.16}$$

where $\Theta_i, i = 1, \ldots, z$ are parameter vectors of suitable dimension, and $\mathbf{\Phi}_i(t)$ is the i-th column of the regressor matrix $\mathbf{\Phi}(t)$.

The above assumption is supported by the following arguments:

- if the zones are homogeneous and/or insulation is properly designed, as it happens in office or government buildings, the heat transfer between neighboring zones is really small;
- numerical simulations show good performance of a decoupled identified model of a well-established test case (see Sect. 3.5);
- possible discrepancies between the model and the real building can be compensated at each iteration by using sensor information and a receding horizon optimization approach.

The need for a receding horizon strategy is further supported by the observation that optimizing over a long time horizon, i.e., one or more days, is quite unreliable. Indeed, satisfying the comfort constraints requires accurate prediction of the indoor temperature of each zone. Such predictions degrade with time due to a number of reasons, like model inaccuracies and lack of reliable weather forecasts. Moreover, long-term energy price forecasts may not be available.

It is worth noticing that if the control signals $u_i(t)$ were continuous rather than binary, then (3.15) would still be a MILP due to the presence of binary DR decision variables u_j^{DR}. However, the computational complexity would be drastically reduced since the number of DR events in each instance of the problem is typically small.

Problem decoupling

In view of Assumption 1, let $\mathbf{U}_i(t, H)$ be the i-th column of $\mathbf{U}(t, H)$, and define the following quantities

$$r_i(t) = w_i u_i(t), \tag{3.17}$$

i.e., the energy consumption of zone Z_i in the time slot t, and

$$R_i(t, H) = \sum_{\tau=t}^{t+H-1} r_i(\tau), \quad C_i(t, H) = \sum_{\tau=t}^{t+H-1} \lambda(\tau) r_i(\tau), \tag{3.18}$$

which amount to total consumption and total cost for Z_i in $\mathcal{I}(t, H)$, respectively. Even when using a decoupled building model, it is apparent that Problem 1 cannot be split into z independent MILPs, one for each zone Z_i, with overall cost function equal to the sum of z marginal costs. Indeed, the decision variables of all zones are coupled through constraint (3.15a) in Problem 1, and the DR component in the cost function itself is a coupling term. In order to overcome this limitation, in the sequel a decoupled optimization problem leading to sub-optimal solutions to Problem 1 is introduced. To this purpose, for each time step t, and for each $\mathcal{R}_j \in \mathcal{P}(t, H)$, the following matrices of real parameters are defined

$$\mathbf{R} = \left\{\overline{r}_{j,i} : i = 1, \ldots, z, \ j \in \mathcal{J}(t, H)\right\}, \tag{3.19}$$

$$\mathbf{\Pi} = \left\{\pi_{j,i} : i = 1, \ldots, z, \ j \in \mathcal{J}(t, H)\right\}, \tag{3.20}$$

where $\{\overline{r}_{j,i}\}$ and $\{\pi_{j,i}\}$ are partitions of $\overline{r}_j$ and π_j, respectively, i.e., $\overline{r}_{j,i} > 0$, $\pi_{j,i} > 0$, and $\overline{r}_j = \sum_{i=1}^{z} \overline{r}_{j,i}$ and $\pi_j = \sum_{i=1}^{z} \pi_{j,i}$. Moreover, the following zone cost is defined:

$$C_i^{\mathcal{P}}(t, H) = C_i(t, H) - \sum_{j \in \mathcal{J}(t,H)} u_{j,i}^{DR} \pi_{j,i}. \tag{3.21}$$

where the new binary variable $u_{j,i}^{DR} \in \mathbb{B}$ takes a different meaning:

$$u_{j,i}^{DR} = \begin{cases} 1 \text{ if zone } i \text{ contributes to the } \mathcal{R}_j \text{ fulfillment} \\ 0 \text{ otherwise.} \end{cases} \tag{3.22}$$

For each $i = 1, \dots z$, consider the following decoupled MILP:

Problem 2 Optimal zone heating control under DR program $\mathcal{P}$.

$$\begin{cases} \mathbf{U}_i^*(t,H) = \arg \min\limits_{\substack{\mathbf{U}_i(t,H) \\ u_{j,i}^{DR} : j \in \mathcal{J}(t,H)}} C_i(t,H) - \sum\limits_{j \in \mathcal{J}(t,H)} u_{j,i}^{DR} \pi_{j,i} \\ \text{s.t.:} \\ R_i(\mu_j, h_j) \leq u_{j,i}^{DR} \overline{r}_{j,i} + (1 - u_{j,i}^{DR}) h_j w_i, \qquad \forall j \in \mathcal{J}(t,H) \\ u_{j,i}^{DR} \in \mathbb{B}, \qquad \forall j \in \mathcal{J}(t,H) \\ T_i^{in}(\tau+1) = \mathbf{\Phi}_i'(\tau) \Theta_i \\ T_i^{in}(\tau) \in \varsigma_i(\tau), \qquad \forall \tau \in \mathcal{I}(t,H) \\ \mathbf{U}_i(t,H) \in \mathbb{B}^{H \times 1} \end{cases} . \tag{3.23}$$

which depends on the particular choice of $\mathbf{R}$ and $\mathbf{\Pi}$. Let $C_i^{*,\mathcal{P}}(t,H)$ be the optimal solution of Problem 2. It is not difficult to see that for any choice of the set of parameters $\mathbf{R}$ and $\mathbf{\Pi}$, the function

$$\overline{C}^{\mathcal{P}}(t,H) = \sum_{i=1}^{z} C_i^{*,\mathcal{P}}(t,H) \tag{3.24}$$

is an upper bound for the optimal cost $C^{*,\mathcal{P}}(t,H)$ of Problem 1. Therefore, it makes sense to look for the values of $\mathbf{R}$ and $\mathbf{\Pi}$ yielding a solution of the z problems (3.23) corresponding to the tightest upper bound, i.e., to find

$$\begin{cases} \min\limits_{\mathbf{R},\mathbf{\Pi}} \overline{C}^{\mathcal{P}}(t,H) \\ \text{s.t.:} \\ \overline{r}_j = \sum_{i=1}^{z} \overline{r}_{j,i}, \quad \pi_j = \sum_{i=1}^{z} \pi_{j,i} \quad \forall j \in \mathcal{J}(t,H) \end{cases} . \tag{3.25}$$

This can be achieved by applying a local constrained minimization algorithm, which involves the solution of z MILPs of the form (3.23) at each step, or via some heuristics. A possible heuristic approach is presented in the following subsection.

Heuristic approach to Problem 2

The heuristics proposed here to assign the parameters $\overline{r}_{j,i}$ and $\pi_{j,i}$ for each zone Z_i and for each DR request $\mathcal{R}_j \in \mathcal{P}(t, H)$ in Problem 2, can be summarized in the following stages:

1. Set the energy price $\lambda(t)$ equal to some big value M for all $t \in \mathcal{I}(\mu_j, h_j)$ $\forall j \in \mathcal{J}(t, H)$, and moreover set $u^{DR}_{j,i} = 0 \ \ \forall j \in \mathcal{J}(t, H), \forall i = 1, \dots, z$. Then solve Problem 2 for each zone Z_i and compute
$$\alpha_{j,i} = \sum_{t \in \mathcal{I}(\mu_j, h_j)} u_i(t) \ \ \forall j \in \mathcal{J}(t, H), \ \ \forall i = 1, \dots, z. \tag{3.26}$$
This step boils down to evaluating the minimum possibile heater activation time $\alpha_{j,i}$ for each zone Z_i within the DR request intervals $\mathcal{I}(\mu_j, h_j)$ in order to satisfy the comfort constraints. The corresponding total energy needed by all zones during $\mathcal{I}(\mu_j, h_j)$ is given by
$$\underline{e} = \sum_{i=1}^{z} \alpha_{j,i} w_i. \tag{3.27}$$
2. If for some j, the quantity $\underline{e}$ exceeds $\overline{r}_j$, then there is no feasible solution to Problem 2 such that the DR request $\mathcal{R}_j$ is satisfied, i.e., under no circumstances the request $\mathcal{R}_j$ can be met without violating comfort constraints. Therefore, for all j such that $\underline{e} > \overline{r}_j$, the request $\mathcal{R}_j$ will be discarded, i.e., $\mathcal{J}(t, H) \leftarrow \mathcal{J}(t, H) \setminus \{j\}$. Otherwise, $\overline{r}_{j,i} = \alpha_{j,i} w_i \ \forall i = 1, \dots, z$. Moreover, the quantity $\overline{r}_j - \underline{e}$, i.e., the estimated amount of excess energy still allowed by $\mathcal{R}_j$ with respect to the minimum possible consumption, is split among all zones according to weights proportional to the corresponding heater power ratings, i.e., $\overline{r}_{j,i}$ is further set to
$$\overline{r}_{j,i} \leftarrow \overline{r}_{j,i} + (\overline{r}_j - \underline{e}) w_i / \sum_{l=1}^{z} w_l. \tag{3.28}$$
3. The parameters $\pi_{j,i}$ are simply assigned by splitting π_j according to weights proportional to the power ratings, i.e.,
$$\pi_{j,i} = \pi_j w_i / \sum_{l=1}^{z} w_l. \tag{3.29}$$
4. Problem 2 is solved with the newly assigned $\overline{r}_{j,i}$, $\pi_{j,i}$ and $\mathcal{J}(t, H)$.

This heuristics can be performed at each step of a receding horizon implementation, as shown below.

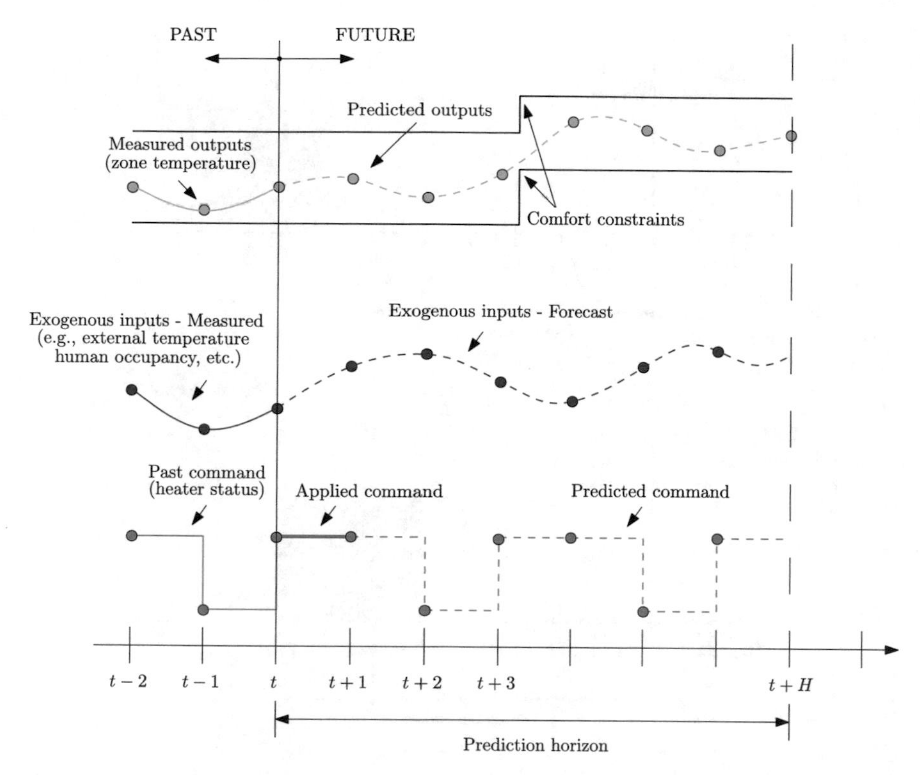

Fig. 3.1 Time evolution of the system variables in standard MPC approach

Receding horizon implementation

Problem 1, as well as the decoupled version in Problem 2 combined with the heuristics just presented, does not lend itself to the standard MPC implementation (see Fig. 3.1), that is,

(i) acquire measurements and/or forecasts of relevant variables,
(ii) optimize the cost over $\mathcal{I}(t, H)$ for fixed H, and
(iii) apply the optimal control action $\mathbf{u}^*(t)$.

Indeed, this basic implementation does not take into account DR requests that partially overlap in time with the moving interval $\mathcal{I}(t, H)$. This issue can be overcome by performing the following further actions right before running the optimization (ii):

(i') adapt the horizon length H dynamically in a way such that an integer number of DR requests falls inside $\mathcal{I}(t, H)$, keeping H greater or equal to a fixed minimum horizon length H_{min} (see Fig. 3.2),
(i") as long as there exists a DR request $\mathcal{R}_j$ such that $\mu_j = t - 1$, set $\mu_j = t$ and reduce h_j by one, subtracting the consumption $r(t - 1)$ [resp. $w_i u_i(t - 1)$] from $\overline{r}_j$ [resp. $\overline{r}_{j,i}$].

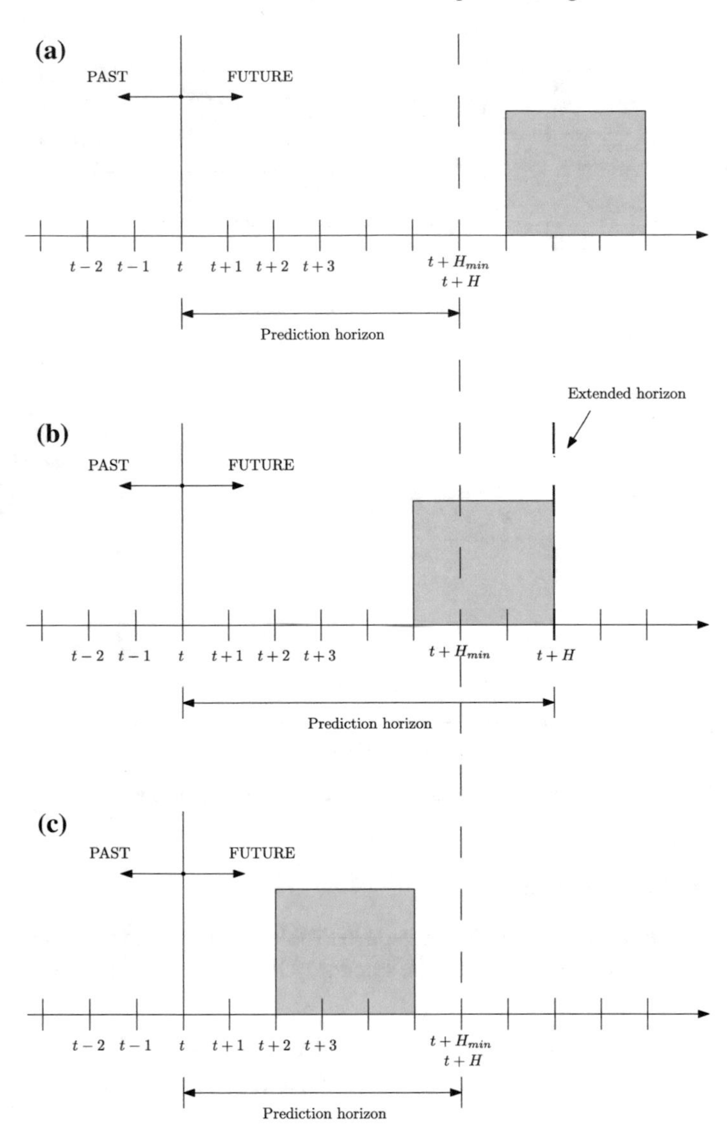

Fig. 3.2 Effect of DR request on the horizon length of the MPC scheme. **a** Standard condition: no DR request within the horizon ($H = H_{min}$). **b** Prediction horizon adaptation: the prediction horizon H is extended to fully cover the incoming DR request ($H > H_{min}$). **c** Standard condition: DR fully contained in the prediction horizon

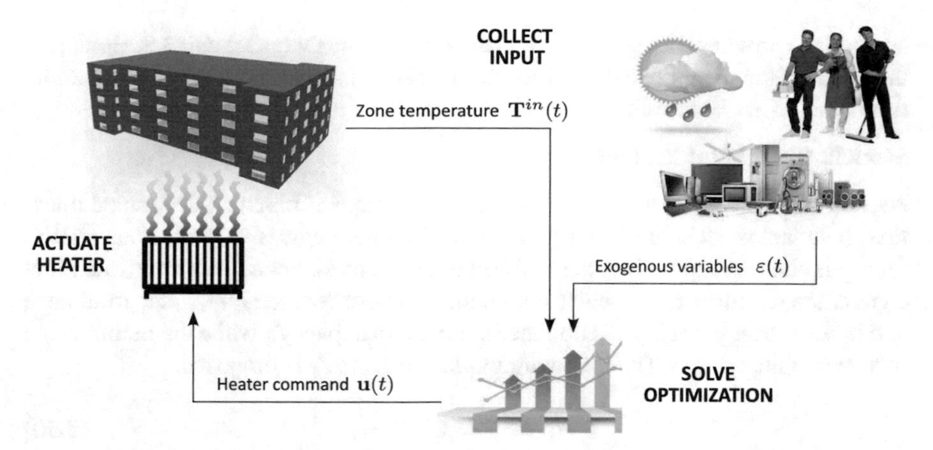

Fig. 3.3 Control system architecture

Algorithm 1 Control Algorithm

for each time step $t = 0, 1, \ldots,$ **do**
  for each zone $i = 1, \ldots, z,$ **do**
    if $t + H_{min}$ falls inside the interval $\mathcal{I}(\mu_j, h_j)$ of the DR request $\mathcal{R}_j$ **then**
      adapt the horizon length H as in step i');
    end if
    if $\mu_j = t - 1$ for some j **then**
      modify the parameters of the DR request $\mathcal{R}_j$ as in step i");
    end if
    acquire regressor data $T_i^{in}(t)$ and $\varepsilon_i(\tau),\ \tau = t, \ldots, t + H - 1$;
    solve Problem 2 for the optimal command sequence $\mathbf{U}_i^*(t, H)$;
    actuate the optimal command action $u_i^*(t)$;
  end for
end for

A schematic of the control architecture is depicted in Fig. 3.3 and the overall procedure is synthesized in Algorithm 1.

3.5 Test Cases

In order to validate the proposed approach, both a small-scale and a large-scale test case involving different heating technologies are developed in this section.

3.5.1 Three-Zone Case

Consider an office building located in Milan (Italy) composed of three zones equipped with radiant floor heating systems. The building characteristics have been taken from an example provided in EnergyPlus and are reported in Tables 3.1 and 3.2. Hereafter, the EnergyPlus model is assumed to be the true (real) building, and the sampling time is set to $\Delta t = 10$ min.

Modeling and identification

As stated in Sect. 3.4.1, the proposed control technique is based on decoupled linear time invariant models of the building zones. The autoregressive exogenous (ARX) family is chosen to model the thermal behavior of zones. For a given zone, the input signals are assumed to be heater command, outdoor temperature, solar irradiance and internal heat gain (lights, appliances, human occupancy), while the output is the indoor air temperature. The following model for zone Z_i is proposed:

$$T_i^{in}(t+1) = \mathbf{\Phi}_i'(t)\mathbf{\Theta}_i, \tag{3.30}$$

Table 3.1 Three-zone case—building features

Building component		Value
Weather and location		Milan
Floor area [m^2]		130.1
Floors [#]		1
Zones [#]		3
Window to wall ratio [%]		0.07
Solar transmittance		0.9
Solar reflectance		0.031
Internal loads	Occupants [#]	10
	Lighting [W/m^2]	1.7
	Equipment [W/m^2]	12.46
Heating: electric low	Power range [kW]	[8; 12]
Temperature radiant system	Throttling range [Δ°C]	2

Table 3.2 Three-zone case—building construction materials (name/thickness [mm])

	External Walls	Internal Walls	Windows
Outside Layer	Cement plaster/25	Gypsum board/10	Generic Clear/3
Layer 2	Concrete block/100	Clay tile/20	
Layer 3	Gypsum board/10	Gypsum board/10	

	Floor	Roof
Outside Layer	Dried sand and gravel/100	Slag or stone/10
Layer 2	Expanded polystyrene/50	Felt and membrane/10
Layer 3	Gypsum concrete/12	Dense insulation/25
Layer 4	Radiant panels/10	HW concrete/50
Layer 4	Gypsum concrete/19	
Layer 5	Tile/2	

being $\boldsymbol{\Theta}_i = \left[\theta_{i,1} \ldots \theta_{i,12}\right]'$, and

$$\begin{aligned}\boldsymbol{\Phi}_i(t) = \big[T_i^{in}(t)\; T_i^{in}(t-1)\; u_i(t)\; u_i(t-1)\; u_i(t-2) \\ G(t)\; G(t-1)\; G(t-2)\; \hat{T}(t+1|t)\; T(t) \\ \hat{I}(t+1|t)\; I(t)\big]',\end{aligned} \tag{3.31}$$

in which $T(t)$ is the outdoor (environment) temperature, $I(t)$ is the external illuminance and $G(t)$ denotes the internal heat gain. $\hat{T}(t+1|t)$ and $\hat{I}(t+1|t)$ denote forecasts of $T(t)$ and $I(t)$ available at time t, respectively.

An identification experiment is conducted to find suitable values for the parameter vectors $\boldsymbol{\Theta}_i$. The identification is performed over two weeks, while validation is done over three different days. The tool used to estimate the parameters of the model from the experiment data is the System Identification Toolbox [32] of Matlab.

For the identification phase, the input signals are chosen as follows:

- the heater command $u_i(t)$ is a pseudo-random binary sequence (PRBS),
- outdoor temperature $T(t)$ and illuminance $I(t)$ are the temperature and illuminance in Milan in January taken from historic time series,
- the internal gain $G(t)$ is a binary signal equal to 1 during the working hours (8:00–18:00) and 0 elsewhere.

Choosing the input as a PRBS in the identification phase is nonetheless standard practice [33] as it allows for persistent excitation of the plant at least in a sufficiently wide frequency range, thus improving the accuracy of the identified model. In Fig. 3.4, a comparison between the real indoor temperature (computed by EnergyPlus) and the 24-step ahead prediction of model (3.30) in the validation days for zone Z_1 is reported. The performance of the estimated model is evaluated by using the *Best Fit* index (FIT) according to the definition in [33, 34]. Roughly speaking, the FIT measures the percentage of signal energy explained by the model. The FIT for all zones turns out to be above 83%. The values of the identified model parameters are reported in Table 3.3.

Experiment setup and discussion

The proposed optimization method is validated on a three-day experiment. In such days, a DR program consisting of 5 DR requests is assumed, as reported in Table 3.4.

Heater power ratings for the three zones are set to 12, 8, and 8 kW, respectively. For all zones, the upper comfort bound is set to $\overline{T}_i = 22$°C throughout the day, while the lower bound is set to $\underline{T}_i = 20$°C from 8:00 to 18:00 and $\underline{T}_i = 16$°C elsewhen. The energy cost profile has been taken from the Italian Electricity Market.

Measured indoor temperature and heater command computed by the proposed method applied to the EnergyPlus physical model are depicted in Fig. 3.5 for zone 1. The overall cost at the computed solution for the three zones is 26.17 €. Not surprisingly, looking at Fig. 3.5 (top), one may observe how the controller tries to preheat a zone before each DR request in order to reduce the power consumption

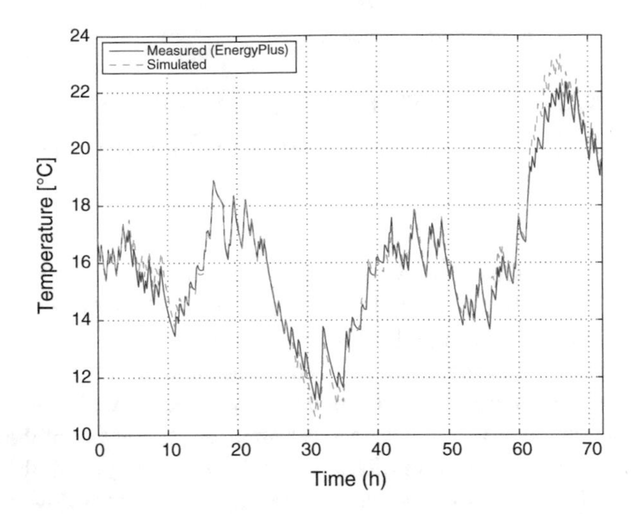

Fig. 3.4 Three-zone case: model validation on 24-step ahead prediction. Comparison between real and predicted output for three validation days (January, 18–20)

Table 3.3 Three-zone case—identified model parameters

	Z_1	Z_2	Z_3
$\theta_{i,1}$	1.307	1.264	1.279
$\theta_{i,2}$	−0.3134	−0.2723	−0.2867
$\theta_{i,3}$	0.7528	0.7575	0.7728
$\theta_{i,4}$	−0.2219	−0.2056	−0.1899
$\theta_{i,5}$	−0.1362	−0.1325	−0.1417
$\theta_{i,6}$	1.06	1.241	1.067
$\theta_{i,7}$	−1.15	−1.283	−1.107
$\theta_{i,8}$	0.1265	0.08702	0.07887
$\theta_{i,9}$	−0.05124	−0.07911	−0.07638
$\theta_{i,10}$	0.0562	0.08536	0.08228
$\theta_{i,11}$	-2.654e-06	-8.778e-06	-3.274e-06
$\theta_{i,12}$	5.263e-06	1.074e-05	4.649e-06

Table 3.4 Three-zone case—DR program

	μ_j	h_j	$\bar{r}_j$ [kWh]	π_j [€]
$\mathcal{R}_1$	67	5	3.9	0.60
$\mathcal{R}_2$	101	3	3.4	0.45
$\mathcal{R}_3$	179	6	4.3	0.75
$\mathcal{R}_4$	325	6	4.2	0.20
$\mathcal{R}_5$	365	6	4.4	0.85

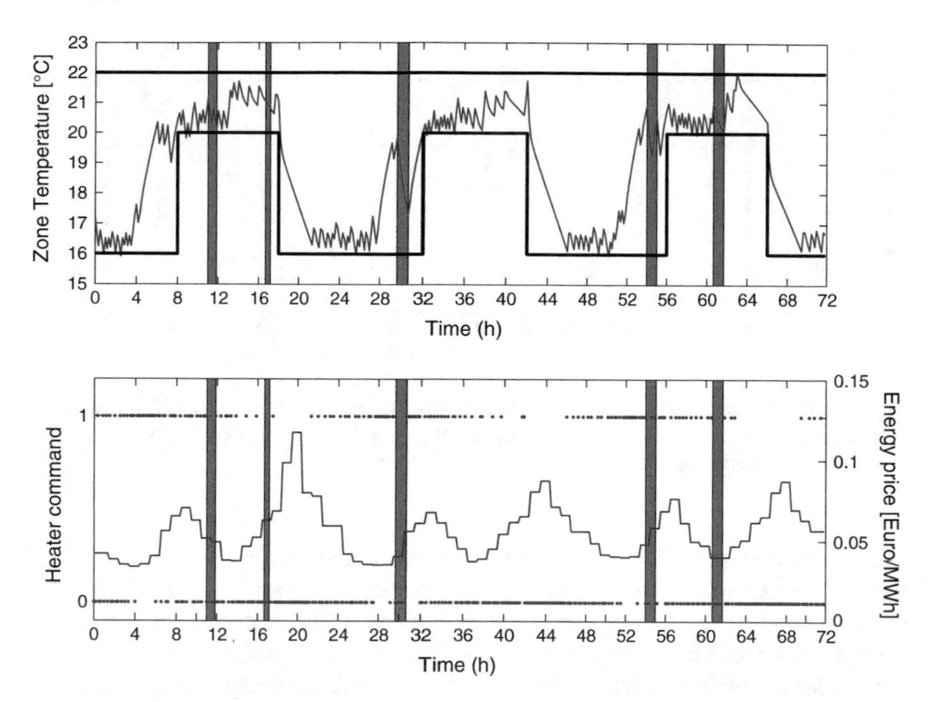

Fig. 3.5 Three-zone case: results obtained through the proposed optimization heuristics on the real scenario (EnergyPlus model). Top: temperature of zone 1 (blue), comfort constraints (black), DR program (green). Bottom: Heater command (blue), energy price (red), DR program (green)

during such a period still maintaining control comfort. To reduce energy cost, the controller also preheats a given zone just before an energy price peak. This fact is easily observable in Fig. 3.5 (bottom), where it is shown that the heater is mainly switched off during on-peak periods. The other two zones exhibit similar behaviors.

In Fig. 3.6, the internal temperature of zone 1 computed on the identified model is reported for both the solutions achieved through the heuristics and the exact optimal algorithm. The zone temperature obtained by both algorithms is quite similar in general. The main differences are due to the different problem solved: while the optimal algorithm computes the solution of a coupled problem (i.e., a problem involving all the building zones), the proposed heuristics works on the decoupled building, i.e., solves one (small) optimization problem for each zone. Although in some places the zone temperature obtained by the two control strategies is different (see, e.g., around hour 62 in the reported simulation), being the building composed of three rooms, temperature differences among zones may compensate giving rise to similar total energy costs. Table 3.5 summarizes the results obtained by running the proposed heuristics and the optimal algorithm on both the real system (EnergyPlus) and the identified model.

By comparing the overall cost obtained by the optimal and the proposed sub-optimal control laws, both implemented on the identified model (see Table 3.5), it

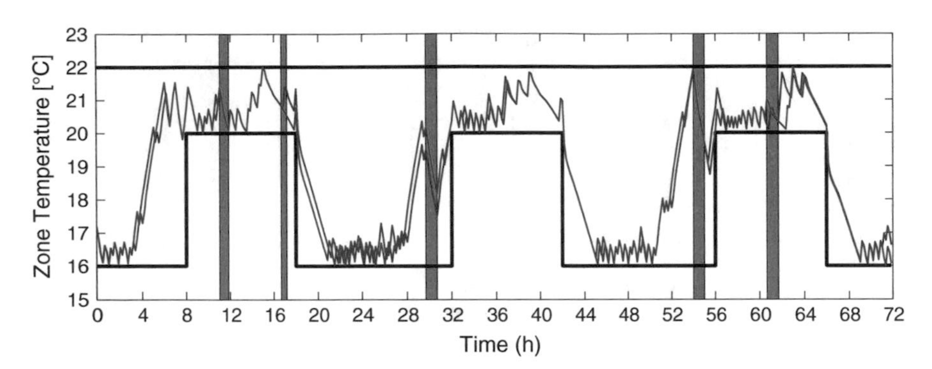

Fig. 3.6 Three-zone case: simulation results obtained on the identified model. Temperature of zone 1 obtained by using the proposed heuristics (blue) and the optimal algorithm (red). Green bands denote DR requests

can be observed that the cost provided by the heuristic suboptimal control is 0.4% higher than that of the optimal control. Unfortunately, it is not possible to obtain an expression for the optimality gap in the general case. The computation time for the suboptimal control law is approximately 2–3 orders of magnitude smaller than that needed for the optimal solution. Actually, as expected, the computational burden of the optimal algorithm, which grows exponentially with the number of zones, leads to intractable problems even when just a few zones, e.g., 4 or 5, are involved. In fact, the computational burden scales in a linear fashion for the heuristic suboptimal algorithm, thus allowing for an efficient solution of large-scale problems, even with hundreds of zones. In addition, it is worthwhile to notice from Fig. 3.6, that the behavior of the controlled variable is very similar for the two alternative control laws.

Concerning the performance obtained in this scenario, it turns out that the proposed suboptimal control law behaves even better than the optimal one (see Table 3.5). Such a behavior is essentially due to noise and modeling errors, and of course this is neither true nor predictable in general. However, by performing further simulations on different data and identified models, the overall costs of both control laws still remain very close. One additional observation concerns the quality of the adopted simplified model. Several identification trials performed on data generated in different conditions by EnergyPlus simulations, invariably show that the decoupled model performs very satisfactorily on validation data. This fact is confirmed quite neatly by the cost achieved by the control law designed on the identified model and applied to the EnergyPlus simulator. The data reported in Table 3.5 show that the degradation of performance is acceptable, being about 3% for the optimal control law and less than 1% for the proposed heuristics.

Table 3.5 Three-zone case - results

	Real [EP]		Simulated [Model]	
	Heur.	Opt.	Heur.	Opt.
Overall cost [€]	26.17	26.86	26.13	26.03
Fulfilled DR reqs	4	4	5	5
DR reward [€]	2.65	2.25	2.85	2.85

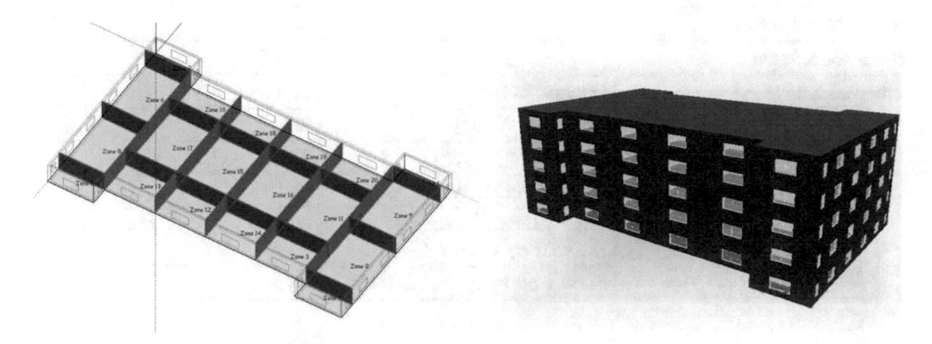

Fig. 3.7 Large-scale case. Map of the first floor and building rendering

3.5.2 Large-Scale Case

To evaluate the behavior of the proposed approach on a more realistic scenario, a 5-floor building composed of 20 zones for each floor is modeled through DesignBuilder, a software tool for developing building models to be used in EnergyPlus simulations. The map of the first floor and a rendered image of the whole building are reported in Fig. 3.7. The building is equipped with heat pump heating systems and its main characteristics are reported in Tables 3.6 and 3.7. The building is still assumed to be located in Milan and the sampling time for simulation is set to 10 min.

Modeling and identification

To apply the proposed heuristics for energy cost reduction, an ARX model for each zone is modeled in a similar way to that used in the three-zone example. A PRBS signal is used as input for estimating the ARX parameters; three days are used for estimation and three for validation. In Fig. 3.8, a comparison between the real output and the 24-step ahead prediction for a given zone is reported. The average FIT for all zones is over 70%.

Experiment setup and discussion

A simulation of three days is performed to evaluate the proposed heuristics. Five DR requests are scheduled in that period as reported in Table 3.8. Comfort profiles and energy cost are set as in Sect. 3.5.1. The application of the proposed method yields an objective value of 235.51 €, including a DR reward of 26.50 €. Figure 3.9 displays

Table 3.6 Large-scale case—building features

Building component		Value
Weather and location		Milan
Floor area [m^2]		1530
Floors [#]		5
Zones [#]		100
Window to wall ratio [%]		30
Solar transmittance		0.84
Solar reflectance		0.075
Internal loads	Occupants [#]	100
	Lighting [W/m^2]	2.36
	Equipment [W/m^2]	1.39
Heating: fan coil unit	Power range [kW]	[3;9]
	Supply humidity ratio [$kgWater/kgDryAir$]	0.0156
	Supply air temperature [°C]	35

Table 3.7 Large-scale case—building construction materials (name/thickness [mm])

	External Walls	**Internal Walls**
Outside Layer	Brickwork/100	Gypsum board/25
Layer 2	Extruded polystyrene/80	Air/10
Layer 3	Concrete block/100	Gypsum board/25
Layer 4	Gypsum plaster/13	

	Floor	**Roof**	**Windows**
Outside Layer	Cast concrete/100	Asphalt/10	Generic Clear/3
Layer 2	Tile/2	Glass wool/140	Air/13
Layer 3		Air/200	Generic Clear/3
Layer 4		Gypsum board/13	

the heater command and the temperature for one zone. The average time needed to compute heater commands for all zones at a given step is less than 20 s.[1] It is worth remarking that one of the novelties of this contribution is the formulation of a relaxed optimization problem based on a decoupled linear regression model of the building. In fact, thanks to decoupling, the computational burden of the proposed algorithm is proportional to the number of zones, thus allowing feasible computations even for large buildings. Moreover, such computations can be easily parallelized with a great reduction of computation time. Like in the previously reported example, the decoupled identified model still shows good performance as depicted in Fig. 3.8.

[1]Computations have been performed using *CPLEX* [35] to solve the LPs, on an Intel Core i5 M520 at 2.40 GHz with 4 GB of RAM.

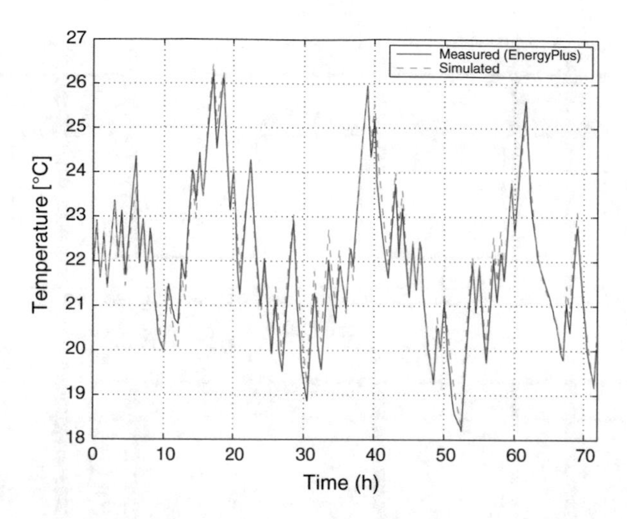

Fig. 3.8 Large-scale case: model validation on 24-step ahead prediction. Comparison between real and predicted output for three validation days (January 25–27)

Table 3.8 Large-scale case—DR program

	h_j	μ_j	$\overline{r}_j$ [kWh]	π_j [€]
$\mathcal{R}_1$	67	5	32	6.00
$\mathcal{R}_2$	101	3	29	4.50
$\mathcal{R}_3$	179	6	41	7.50
$\mathcal{R}_4$	325	6	30	2.00
$\mathcal{R}_5$	365	6	37	8.50

This is an essential prerequisite for the use of MPC techniques. Regarding internal temperature reported in Fig. 3.9, one may notice similarities w.r.t. to the three-zone test case. In particular, preheating before a DR request as well as reduction of power consumption during on-peak times are again observed. In addition, one may notice a fast rising edge of zone temperature before 8 a.m. of each day (i.e., before hour 8, 32 and 56 in the simulation) due to the change in the lower comfort bound. By looking at the DR requests, it becomes apparent how the proposed technique tries to honour such requests by reducing the power consumption in the respective time intervals. However, it is worthwhile to note that in general not all DR requests will be satisfied, but only those that are economically convenient and feasible from the comfort point of view.

As opposed to the three-zone case, it is not possible to compare the suboptimal strategy with the optimal one, due to the numerical intractability of the optimal MILP problem. In order to arrange an experiment to evaluate the quality of the proposed method, the binary commands of the heaters are relaxed to be continuous, i.e., it

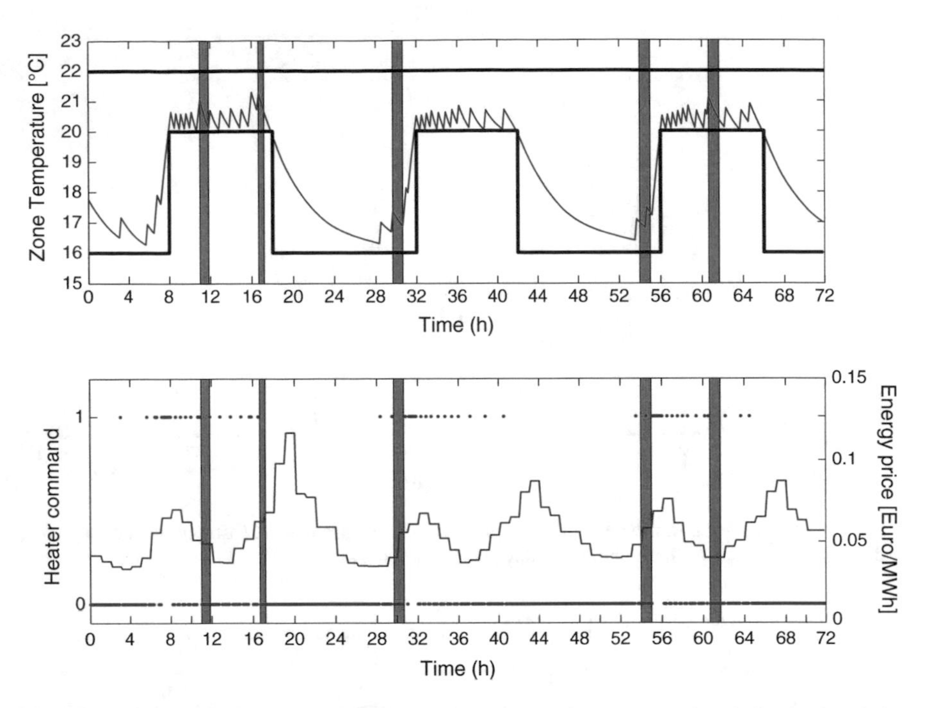

Fig. 3.9 Large-scale case: simulation results obtained through the proposed optimization heuristics on the identified model. Top: temperature of zone 20 (blue), comfort constraints (black), DR program (green). Bottom: Heater command (blue), energy price (red), DR program (green)

is assumed that heaters may change continuously their power from zero to their maximum. Although the considered scenario may not seem feasible from a realistic point of view, it turns out to be convenient for estimating the quality of performance of the proposed approach. In fact, the obtained results provide a lower bound to the optimal cost. The total costs and the DR rewards for both optimizations are reported in Table 3.9. Notice that the cost obtained by the proposed method turns out to be about 4% greater w.r.t. the relaxed optimal one. It is reasonable to expect that the devised heuristics will behave similarly in the original (non relaxed) problem, too. In Fig. 3.10, the behavior of the internal temperature of zone 20 for both strategies is reported. Notice that the temperature profiles are almost indistinguishable except in the proximity of the first and the fourth DR requests. As reported in Table 3.9, the proposed heuristics is able to fulfill 3 DR requests out of 5, contrary to the optimal algorithm which is able to honor all the DR requests. In any case, as previously stated, the gap of the proposed heuristics is just 4% greater w.r.t. the optimal one, showing good performance of the method.

Table 3.9 Large-scale case—Overall simulation results obtained by assuming continuous regulation of heaters' power (relaxed model)

	Relaxed model—simulated	
	Heuristic	Optimal
Overall cost [€]	200.39	192.47
Fulfilled DR reqs	3	5
DR reward [€]	20.50	28.50

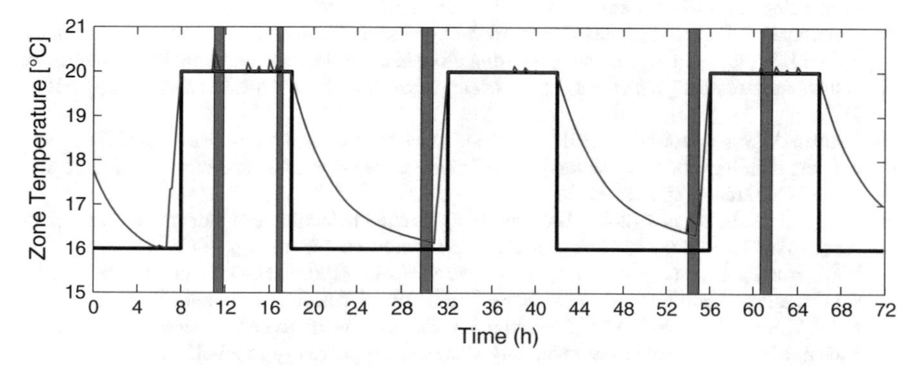

Fig. 3.10 Large-scale case: simulation results obtained by assuming continuous regulation of heaters power (relaxed model). Temperature of zone 20 obtained by using the proposed heuristics (blue) and the optimal algorithm (red). Green bands denote DR requests

3.6 Conclusion

In this chapter the problem of optimizing the operation of a building heating system under the hypothesis of participation in a DR program has been addressed. The DR setup is based on price-volume signals sent by an aggregator to the building energy management system. The optimizer exploits a receding horizon control technique for minimizing the energy bill. Since the complexity of the overall optimization is intractable even for buildings of modest dimension, a heuristic search strategy based on problem decomposition has been devised to make the computational burden reasonable. Numerical results show that the heuristic strategy involves almost negligible loss of accuracy with respect to the exact optimal solution. The overall optimization procedure has been tested both on the simplified identified model used for design, and on a realistic building model computed using EnergyPlus. The results show excellent performance in terms of robustness of the control law to model uncertainties.

References

1. Bianchini G, Casini M, Vicino A, Zarrilli D (2014) Receding horizon control for demand-response operation of building heating systems. In: Proceedings of IEEE conference on decision and control, pp 4862–4867
2. Bianchini G, Casini M, Vicino A, Zarrilli D (2016) Demand-response in building heating systems: a model predictive control approach. Appl Energy 168:159–170
3. Oldewurtel F, Ulbig A, Parisio A, Andersson G, Morari M (2010) Reducing peak electricity demand in building climate control using real-time pricing and model predictive control. In: Proceedings of IEEE conference on decision and control, pp 1927–1932
4. Oldewurtel F, Parisio A, Jones C, Morari M, Gyalistras D, Gwerder M, Stauch V, Lehmann B, Wirth K (2010) Energy efficient building climate control using stochastic model predictive control and weather predictions. In: Proceedings of American control conference, pp 5100–5105
5. Kelman A, Ma Y, Borrelli F (2011) Analysis of local optima in predictive control for energy efficient buildings. In: Proceedings of IEEE conference on decision and control and European control conference, pp 5125–5130
6. Cole W, Hale E, Edgar T (2013) Building energy model reduction for model predictive control using OpenStudio. In: Proceedings of American control conference, pp 449–454
7. Ma Y, Borrelli F, Hencey B, Coffey B, Bengea S, Haves P (2012) Model predictive control for the operation of building cooling systems. IEEE Trans Control Syst Technol 20(3):796–803
8. Kwak Y, Huh J-H, Jang C (2015) Development of a model predictive control framework through real-time building energy management system data. Appl Energy 155:1–13
9. Losi A, Mancarella P, Vicino A (2015) Integration of demand response into the electricity chain: challenges, opportunities and smart grid solutions. Wiley
10. Zhou Z, Zhao F, Wang J (2011) Agent-based electricity market simulation with demand response from commercial buildings. IEEE Trans Smart Grid 2(4):580–588
11. Hong SH, Yu M, Huang X (2015) A real-time demand response algorithm for heterogeneous devices in buildings and homes. Energy 80:123–132
12. Kelly GE (1988) Control system simulation in North America. Energy Build 10(3):193–202
13. Bilgin E, Caramanis MC, Paschalidis IC (2013) Smart building real time pricing for offering load-side regulation service reserves. In: Proceedings of IEEE conference on decision and control, pp 4341–4348
14. Coffey B, Haghighat F, Morofsky E, Kutrowski E (2010) A software framework for model predictive control with GenOpt. Energy Build 42(7):1084–1092
15. Xue X, Wang S, Yan C, Cui B (2015) A fast chiller power demand response control strategy for buildings connected to smart grid. Appl Energy 137:77–87
16. Xue X, Wang S, Yan C, Cui B (2014) Dynamic demand response controller based on real-time retail price for residential buildings. IEEE Trans Smart Grid 5(1):121–129
17. Yoon JH, Baldick R, Novoselac A (2014) Demand response for residential buildings based on dynamic price of electricity. Energy Build 80:531–541
18. Crawley DB, Pedersen CO, Lawrie LK, Winkelmann FC (2000) Energyplus: energy simulation program. ASHRAE J 42(4):49
19. Nghiem T, Behl M, Mangharam R, Pappas G (2011) Green scheduling of control systems for peak demand reduction. In: Proceedings of IEEE conference on decision and control and European control conference, pp 5131–5136
20. Nghiem T, Behl M, Mangharam R, Pappas G (2012) Scalable scheduling of building control systems for peak demand reduction. In: Proceedings of American Control Conference, pp 3050–3055
21. Nghiem T, Pappas G, Mangharam R (2013) Event-based green scheduling of radiant systems in buildings. In: Proceedings of American control conference, pp 455–460
22. Albadi MH, El-Saadany E (2007) Demand response in electricity markets: An overview. In: Proceedings of IEEE PES general meeting, pp 1–5

23. Baboli P, Moghaddam M, Eghbal M (2011) Present status and future trends in enabling demand response programs. In: Proceedings of IEEE PES general meeting, pp 1–6
24. Balijepalli VM, Pradhan V, Khaparde S, Shereef R, (2011) Review of demand response under smart grid paradigm. In: Proceedings of IEEE PES innovative smart grid technologies Indian conference, pp 236–243
25. Belhomme R, Cerero Real de Asua R, Valtorta G, Paice A, Bouffard F, Rooth R, Losi A (2008) ADDRESS—active demand for the smart grids of the future. In: Proceedings of CIRED seminar: smart grids for distribution
26. Belhomme R, Cerero Real de Asua R, Valtorta G, Eyrolles, P (2011) The ADDRESS project: developing active demand in smart power systems integrating renewables. In: Proceedings of IEEE PES general meeting
27. Koponen P, Ikaheimo J, Vicino A, Agnetis A, De Pascale G, Ruiz Carames N, Jimeno J, Sanchez-Ubeda E, Garcia-Gonzalez P, Cossent R (2012) Toolbox for aggregator of flexible demand. In: Proceedings of IEEE international energy conference and exhibition, pp 623–628
28. Paoletti S, Vicino A, Zarrilli D (2015) Computational methods for technical validation of demand response products. In: Proceedings of IEEE international conference on environment and electrical engineering, pp 117–122
29. Chassin DP, Stoustrup J, Agathoklis P, Djilali N (2015) A new thermostat for real-time price demand response: cost, comfort and energy impacts of discrete-time control without deadband. Appl Energy 155:816–825
30. Patteeuw D, Bruninx K, Arteconi A, Delarue E, D'haeseleer W, Helsen L (2015) Integrated modeling of active demand response with electric heating systems coupled to thermal energy storage systems. Appl Energy 151:306–319
31. Cardell J, Anderson C (2015) Targeting existing power plants: EPA emission reduction with wind and demand response. Energy Policy 80:11–23
32. The MathWorks system identification toolbox. http://www.mathworks.com/products/sysid
33. Ljung L (1999) System identification: theory for the user, 2nd edn. Prentice Hall PTR, Upper Saddle River
34. Ljung L (2007) System identification toolbox for use with MATLAB: user's guide. Natick
35. IBM, IBM ILOG Cplex optimizer. http://www-01.ibm.com/software/integration/optimization/cplex-optimizer/

Chapter 4
Configuration and Control of Storages in Distribution Networks

In this chapter, the potential of applying ESS for facilitating the integration of renewables into electric power systems is presented. In particular, the possibility of using ESSs for mitigating over- and undervoltages in LV networks is investigated. Moreover, different algorithms for both planning and operation of such innovative devices are proposed. The material of this chapter is mainly based on [1–4].

4.1 Introduction

The growing penetration of low carbon technologies, such as distributed generation (DG), electric vehicles and heat pumps, is determining significant modifications of typical power flows in electrical grids. One example is the switching of distribution networks from "passive" to "active" and vice versa (power flow inversion), which causes serious problems at the interface with the transmission network. Another related issue is represented by voltage problems. Since peaks of load demand and DG are typically not aligned in time, over- and undervoltage conditions may regularly show up in LV networks with higher frequency. This implies a progressive degradation of the quality of service provided to consumers. Therefore, maintaining voltage between specified limits, as required by typical quality of supply standards, has become one of the main issues currently faced by DSO. Traditional grid reinforcement, which consists of replacing existing cables and/or transformers with new equipments, is a possible solution to mitigate these problems. However, in view of a dynamic scenario, in which new loads and/or generators might be continuously connected to or disconnected from the distribution network, more flexible and/or cheaper solutions could be considered among those presented in the literature [5]. Initial investigations on the use of on-load tap changers (OLTC) at secondary substations are available, see, e.g., [6] and references therein. In [7], different voltage control schemes involving active and reactive power control of DG inverters are compared.

D. Zarrilli, *Integration of Low Carbon Technologies in Smart Grids*,
Springer Theses, https://doi.org/10.1007/978-3-319-98358-5_4

However, these solutions cannot be always put into practice. In fact, depending on both technical and regulatory issues, the DSO might not have control over distributed generators to directly operate adjustable inverters. On the other hand, since secondary substations are generally not automated, OLTC control becomes impracticable. Moreover, should voltage rise and drop occur simultaneously at different buses of the network, voltage control at the secondary substation level is ineffective. For the case when feeders show contrasting voltage issues, in [8] the authors investigate how distributed capacitor banks can help provide additional flexibility to the OLTC solution. In addition, this type of control might have the serious drawback to transmit large disturbances to the medium voltage (MV) network.

An alternative to the aforementioned solutions, which has recently received increasing attention, is represented by ESSs, see, e.g., [9–11] and references therein. The idea is that an energy storage should play the role of a load in case of overvoltage and that of a generator in case of undervoltage. Pros of the use of ESSs are that voltage problems are solved locally, thus limiting the impact on the MV network, and curtailment of renewable generation is minimized. These advantages add to the other benefits that the use of ESS brings to the different power system stakeholders (see, e.g., [12–15] for a general survey).

ESS deployment in power networks involves problems at both the planning and operation levels. In the planning stage, the number of ESSs to be installed, their locations and sizes must be decided. Optimal ESS siting and sizing procedures have been proposed in the literature for both transmission and distribution networks, and from the point of view of different power system stakeholders (see, e.g., [16–19] and references therein). A quite general approach is to formulate the problem in an OPF framework. Storage locations and sizes are considered as optimization variables, and a cost function (including, e.g., generation costs, storage installation costs, network losses, etc.) is minimized, subject to power flow constraints and storage dynamics. Since integer variables used to decide where ESSs should be installed and non-convex power flow constraints make the resulting optimization problems NP-hard, different approaches have been devised to approximate the exact problems and alleviate the computational burden. Linearization via DC approximation is adopted when dealing with transmission networks, where the assumptions underlying DC OPF are typically valid [20–23]. When the full AC OPF is considered, as is common for distribution networks, appropriate convex relaxations are often exploited. The second-order cone programming OPF approach of [24] is considered in [25], while convex relaxations based on SDP are used in [26–28]. Alternating direction method of multipliers is proposed in [11] to break down the original problem into a distributed parallel convex optimization. In [25, 28] convex relaxations are embedded in a mixed integer formulation of the siting problem. In [16], the computationally demanding unit commitment (UC) problem over one year to determine optimal storage locations and parameters, is tackled by solving a UC problem for each day of the year separately.

Conversely, operation is concerned with the use of the installed ESSs, and requires the design of a control policy according to which the ESSs are operated to guarantee the satisfaction of the required specifications. The definition of effective ESS control policies for voltage control in LV networks struggles with the poor availability of measurements, which makes reconstruction and forecasting of the state of the network a challenging problem. For this reason, several approaches proposed in the literature address voltage control under different partial information setups. In [9], a coordinated control scheme is proposed where a centralized controller decides which ESSs are to be activated upon the occurrence of a voltage problem, while dedicated decentralized controllers operate each ESS having visibility only of local voltage measurements. In [29], the performance of an OLTC control logic is assessed under three different remote monitoring schemes, namely measurements taken at the middle, end, or middle and end of each feeder. Though these approaches are appealing because computationally cheap, one limitation is their lack of predictive capability, which may turn out into sending the control signal to ESSs with delay, and without the guarantee that ESSs are ready for the required actions. In [9], where the ESSs are used to solve daytime overvoltages due to PV generation, this problem is tackled by discharging the ESSs overnight. Apparently, this solution is not viable if over- and undervoltages occur irregularly.

In this chapter, the whole process of integrating ESSs in a distribution network for voltage control purposes, from the deployment to their optimal control, is analyzed. In the considered framework, the stakeholder is assumed to be the DSO, which faces regularly over- and/or undervoltages in a LV distribution network, and looks for solutions to this issue in order to avoid penalties imposed by the regulatory framework. Consequently, ESSs are assumed to be installed and operated by the DSO, i.e., they are not integrated in households or DG plants out of the control of the DSO. As far as the configuration stage is concerned, a voltage sensitivity analysis has been proposed to circumvent the combinatorial nature of the siting problem and a multi-period OPF has been adopted to determine the size of each storage unit. Moreover, a control algorithm based on the receding horizon control framework has been devised to optimal operate ESSs and anticipate possible voltage problems. Applications presented in the last part of the chapter (a real Italian LV network, a modified version of the IEEE 34-bus test feeder, and 200 randomly generated radial networks) provide encouraging results when the ESS allocation strategy is compared with exhaustive search. They bring interesting insights in how the network topology, and historical load demand and DG profiles, can be exploited to identify the most suitable candidate buses for ESS siting. Furthermore, simulations show how the proposed ESS control policy could greatly enhance the voltage quality at customers' premises.

The chapter is organized as follows. Section 4.2 presents the network model. Section 4.3 introduces the OPF problem and its convex relaxation. Both the sizing and the siting problem are addressed in Sect. 4.4, whereas the proposed algorithm for real-time ESS control is described in Sect. 4.5. Experimental results are reported and discussed in Sect. 4.6. Finally, conclusions are drawn in Sect. 4.7.

4.2 Network Model

Consider a radial LV distribution network. The network is described by a graph $(\mathcal{N}, \mathcal{E})$, where $\mathcal{N} = \{1, 2, \ldots, n\}$ is the set of nodes (*buses*) and $\mathcal{E}$ is the set of edges (*lines*). Bus 1 is assumed to be the slack bus, representing the interconnection with the MV network, while the set $\mathcal{N}^L = \{2, \ldots, n\}$ includes all the remaining buses. Moreover, the set of buses having generation is denoted by $\mathcal{G} \subseteq \mathcal{N}^L$. The admittance-to-ground at bus i is denoted by y_{ii}, while the line admittance between nodes i and j is denoted by y_{ij}. Obviously, $y_{ij} = y_{ji}$. If $(i, j) \notin \mathcal{E}$, i.e., buses i and j are not connected by a line, $y_{ij} = 0$. The network admittance matrix $\boldsymbol{Y} = [Y_{ij}] \in \mathbb{C}^{n\times n}$ is a symmetric matrix defined as

$$Y_{ij} = \begin{cases} y_{ii} + \sum_{h\neq i} y_{ih} & \text{if } i = j \\ -y_{ij} & \text{otherwise.} \end{cases} \tag{4.1}$$

Consider discrete time steps denoted by $t = 1, 2, \ldots$. The complex voltage at bus k and time t is denoted by $V_k(t)$. The slack bus is characterized by fixed voltage magnitude and phase. Voltage magnitude and phase at all other buses in the set $\mathcal{N}^L$ are determined by the network operating conditions. For $k \in \mathcal{N}^L$, voltage quality requirements impose the voltage magnitude to remain within specified limits, i.e.,

$$\underline{v}_k^2 \leq |V_k(t)|^2 \leq \overline{v}_k^2, \tag{4.2}$$

where $\underline{v}_k \leq \overline{v}_k$ are given bounds. Moreover, real power flow from bus i to bus j is bounded according to the physical properties of the lines. This implies the constraint

$$p_{ij}(t) \leq \overline{p}_{ij}, \tag{4.3}$$

where $p_{ij}(t) = \text{Re}\big(V_i(t)\big[V_i(t) - V_j(t)\big]^* y_{ij}^*\big)$ is the real power transferred from bus i to bus j at time t, and $\overline{p}_{ij} = \overline{p}_{ji}$ is a given upper bound depending only on the physical properties of the line [30]. Denoting the active and reactive power injections at bus k by p_k and q_k, respectively, the power balance equations at bus k and time t read as

$$p_k(t) = \text{Re}\Big(V_k(t) \sum_{j\in\mathcal{N}} V_j^*(t) Y_{kj}^*\Big) \tag{4.4a}$$

$$q_k(t) = \text{Im}\Big(V_k(t) \sum_{j\in\mathcal{N}} V_j^*(t) Y_{kj}^*\Big). \tag{4.4b}$$

In turn, for a bus k having loads and generators connected to it, p_k and q_k can be decomposed as

$$p_k(t) = p_k^{res}(t) - p_k^{load}(t) \tag{4.5a}$$

$$q_k(t) = q_k^{res}(t) - q_k^{load}(t), \tag{4.5b}$$

where the superscript ^{res} refers to generation and the superscript ^{load} refers to demand. By putting together (4.4) and (4.5), it follows

$$\mathrm{Re}\Big(V_k(t)\sum_{j\in\mathcal{N}} V_j^*(t)Y_{kj}^*\Big) = p_k^{res}(t) - p_k^{load}(t) \tag{4.6a}$$

$$\mathrm{Im}\Big(V_k(t)\sum_{j\in\mathcal{N}} V_j^*(t)Y_{kj}^*\Big) = q_k^{res}(t) - q_k^{load}(t). \tag{4.6b}$$

Active power $p_1(t)$ and reactive power $q_1(t)$ injected at the slack bus are determined by the power flow. For all other buses in the set $\mathcal{N}^L$, $V_k(t)$ are free variables of the power flow problem, while the quantities $p_k^{load}(t)$, $q_k^{load}(t)$, $p_k^{res}(t)$ and $q_k^{res}(t)$ are considered as known inputs in (4.6). In case no load or generator is connected to bus k, the corresponding demand or generation are assumed to be zero, i.e., $p_k^{res}(t) = q_k^{res}(t) = 0, \quad \forall k \in \mathcal{N}^L \backslash \mathcal{G}$.

4.3 OPF and SDP Relaxation

In a typical OPF problem, cost function includes the cost of generating real power or line losses over the network. Note that all these costs can be expressed as functions of $\boldsymbol{V}(t) = [\ V_1(t)\ \ldots\ V_n(t)\ ]'$. Let the cost function be $C\big(\boldsymbol{V}(t)\big)$. Then the OPF of interest can be formulated as

Problem 3 Standard OPF problem.

$$\begin{cases} \min\limits_{\boldsymbol{V}(t)} C\big(\boldsymbol{V}(t)\big) \\ \text{s.t.:} \\ (4.2) - (4.6), \quad k \in \mathcal{N}^L, \quad (i,j) \in \mathcal{E} \end{cases}. \tag{4.7}$$

Unfortunately, Problem 3 is non-convex (due to the nonconvexity of the constraint (4.2) and the bilinearity in the equality constraint in (4.6)), and NP-hard to solve in general. A common approach to reduce the computational burden is to compute an approximated solution by resorting to SDP convex relaxations [31]. This approach suggests using a new variable $\boldsymbol{W}(t) = \boldsymbol{U}(t)\boldsymbol{U}(t)'$ as the main voltage information, where $\boldsymbol{U}(t) = \big[\ \mathrm{Re}\big(\boldsymbol{V}(t)\big)'\ \mathrm{Im}\big(\boldsymbol{V}(t)\big)'\ \big]'$ is the decomposed voltage vector. Let ϵ_k be the k-th standard basis vector of R^n, and let

$$\mathcal{Y}_k = \epsilon_k \epsilon_k' Y \tag{4.8}$$

$$\mathcal{Y}_{ij} = y_{ij}\epsilon_i(\epsilon_i - \epsilon_j)' \tag{4.9}$$

$$\boldsymbol{Y}_k^p = \frac{1}{2}\begin{bmatrix} \mathrm{Re}\big(\mathcal{Y}_k + \mathcal{Y}_k'\big) & \mathrm{Im}\big(\mathcal{Y}_k' - \mathcal{Y}_k\big) \\ \mathrm{Im}\big(\mathcal{Y}_k - \mathcal{Y}_k'\big) & \mathrm{Re}\big(\mathcal{Y}_k + \mathcal{Y}_k'\big) \end{bmatrix} \tag{4.10}$$

$$\boldsymbol{Y}_k^q = \frac{-1}{2}\begin{bmatrix} \mathrm{Im}(\mathcal{Y}_k + \mathcal{Y}_k') & \mathrm{Re}(\mathcal{Y}_k - \mathcal{Y}_k') \\ \mathrm{Re}(\mathcal{Y}_k' - \mathcal{Y}_k) & \mathrm{Im}(\mathcal{Y}_k + \mathcal{Y}_k') \end{bmatrix} \tag{4.11}$$

$$\boldsymbol{Y}_{ij}^p = \frac{1}{2}\begin{bmatrix} \mathrm{Re}(\mathcal{Y}_{ij} + \mathcal{Y}_{ij}') & \mathrm{Im}(\mathcal{Y}_{ij}' - \mathcal{Y}_{ij}) \\ \mathrm{Im}(\mathcal{Y}_{ij} - \mathcal{Y}_{ij}') & \mathrm{Re}(\mathcal{Y}_{ij} + \mathcal{Y}_{ij}') \end{bmatrix} \tag{4.12}$$

$$\boldsymbol{M}_k = \begin{bmatrix} \epsilon_k \epsilon_k' & 0 \\ 0 & \epsilon_k \epsilon_k' \end{bmatrix}. \tag{4.13}$$

It can be easily verified that with the previous definition

$$|V_k(t)|^2 = \mathrm{Tr}\big(\boldsymbol{M}_k \boldsymbol{W}(t)\big) \tag{4.14a}$$

$$p_{ij}(t) = \mathrm{Tr}\big(\boldsymbol{Y}_{ij}^p \boldsymbol{W}(t)\big) \tag{4.14b}$$

$$p_k(t) = \mathrm{Tr}\big(\boldsymbol{Y}_k^p \boldsymbol{W}(t)\big) \tag{4.14c}$$

$$q_k(t) = \mathrm{Tr}\big(\boldsymbol{Y}_k^q \boldsymbol{W}(t)\big). \tag{4.14d}$$

The interested reader is referred to [30] for a detailed derivation of (4.14). Adding all together, one can show that Problem 3 is equivalent to the following formulation.

Problem 4 SDP OPF problem.

$$\begin{cases} \min\limits_{\boldsymbol{W}(t)} C\big(\boldsymbol{W}(t)\big) \\ \text{s.t.:} \\ \\ \mathrm{Tr}\big(\boldsymbol{Y}_k^p \boldsymbol{W}(t)\big) = p_k^{res}(t) - p_k^{load}(t) \\ \mathrm{Tr}\big(\boldsymbol{Y}_k^q \boldsymbol{W}(t)\big) = q_k^{res}(t) - q_k^{load}(t) \\ \underline{v}_k^2 \le \mathrm{Tr}\big(\boldsymbol{M}_k \boldsymbol{W}(t)\big) \le \overline{v}_k^2 \\ \mathrm{Tr}\big(\boldsymbol{Y}_{ij}^p \boldsymbol{W}(t)\big) \le \overline{p}_{ij} \\ \\ \mathrm{rank}\big(\boldsymbol{W}(t)\big) = 1 \\ \boldsymbol{W}(t) \succeq 0 \\ k \in \mathcal{N}, \quad (i,j) \in \mathcal{E} \end{cases} . \tag{4.15}$$

The equivalence between (4.7) and (4.15) follows from the fact that the symmetric matrix $\boldsymbol{W}(t) \in \mathbb{R}^{n \times n}$ is positive semidefinite and of rank-1 if and only if there exists $\boldsymbol{U}(t) \in \mathbb{R}^n$ such that $\boldsymbol{W}(t) = \boldsymbol{U}(t)\boldsymbol{U}(t)'$. Assume that the cost function is convex, as it is frequent in OPF formulations, and move our attention to the feasibility set only. Unlike (4.7) which is quadratic in $\boldsymbol{V}(t)$, this problem is convex in $\boldsymbol{W}(t)$ except the nonconvex rank-1 constraint. Removing the rank-1 constraint yields the standard SDP relaxation. In general, the relaxation provides a lower bound to the optimal cost of the OPF problem (4.7), but it is shown empirically in [32] that it works particularly well for radial distribution networks. Sufficient conditions under which the relaxation is exact, are provided in [33].

4.4 ESS Configuration

In this section, the problem of finding the optimal number, location and size of ESSs with the aim of preventing over- and undervoltages in a radial LV network is considered. The combinatorial nature of the siting problem is coped with by devising a heuristic strategy based on voltage sensitivity analysis. The heuristic determines the location of a given number of ESSs in the network that are expected to be most effective for voltage support. Then, for fixed storage locations, ESS sizes are determined by formulating a multi-period OPF problem [27, 34], which is then tackled by resorting to SDP convex relaxations [31]. The final choice of the best ESS allocation is done by minimizing a total cost, which takes into account the number of storage devices, their total installed capacity, and the average network losses. Following the taxonomy in [17], the contribution of the ESS configuration step can be positioned as a mix of heuristic and mathematical programming methods. Its unique feature in the framework of the related literature is the siting procedure, which exploits network topology and line parameters. Advantages of the proposed placement procedure are twofold. First, it is computationally cheap and avoids resorting to a combinatorial formulation involving integer variables. Second, it is specifically designed in order to ease the solution of the subsequent sizing problem by selecting locations with highest sensitivity to solve over- and undervoltages. Similarly to [16], uncertainties in the storage sizing decision problem, mainly arising from stochastic generation and demand, are taken into account by carrying out the optimal sizing over different realizations of the demand and generation profiles. For each installed storage, the largest size over the considered time horizon is finally selected. Roughly speaking, this "worst-case" approach is motivated by the attempt to "robustify" the decision by maximizing the probability of feasible storage operation under all possible realizations of demand and generation. Overall, the goal of the proposed procedure is to find a cost-effective storage installation strategy providing voltage support to the LV network, while keeping the computational burden affordable.

ESS model

Assume that distributed energy storage can be installed in the network, and let $\mathcal{S} \subseteq \mathcal{N}^L$ be the set of buses equipped with ESS. For $s \in \mathcal{S}$, $e_s(t)$ denotes the storage energy level at bus s and time t. The dynamics of $e_s(t)$ is modeled to follow the first-order difference equation

$$e_s(t) = e_s(t-1) + p_s^{ess}(t-1)\Delta t, \tag{4.16}$$

where $p_s^{ess}(t-1)$ is the average active power pumped into ($p_s^{ess} > 0$) or drawn from ($p_s^{ess} < 0$) the storage between $t-1$ and t, and Δt is the time step. The initial condition for the storage level is assumed to be known

$$e_s(1) = e_s^{start}. \tag{4.17}$$

Moreover, both $p_s^{ess}(t-1)$ and $e_s(t)$ are bounded as follows

$$\underline{p}_s^{ess} \leq p_s^{ess}(t-1) \leq \overline{p}_s^{ess} \tag{4.18}$$
$$0 \leq e_s(t) \leq \overline{e}_s, \tag{4.19}$$

where $\underline{p}_s^{ess} < 0$ and $\overline{p}_s^{ess} > 0$ are the ramp rate limits, and $\overline{e}_s$ is the storage capacity installed at bus s. Similarly to (4.18), the average reactive power $q_s^{ess}(t-1)$ exchanged by the compensation device between $t-1$ and t can be constrained by imposing

$$\underline{q}_s^{ess} \leq q_s^{ess}(t-1) \leq \overline{q}_s^{ess}, \tag{4.20}$$

for fixed bounds $\underline{q}_s^{ess} < \overline{q}_s^{ess}$.

Problem formulation

The configuration cost of distributed storage systems in the grid depends both on the number of devices (e.g., costs incurred for installation and maintenance) and the total amount of energy storage capacity installed. Moreover, the effectiveness of a given ESS in providing voltage regulation services can vary significantly, depending on the bus to which it is connected. Hence, the following problem need be tackled at the planning stage.

Problem 5 Optimal ESS configuration problem: Find the minimum-cost ESS *number, location* and *size* for preventing over- and undervoltages in the considered LV network.

Solving Problem 5 is a formidable task, since it requires to solve a multi-period OPF, with mixed binary and continuous optimization variables. In order to circumvent the combinatorial nature of the problem, and reduce its computational complexity, a two-step, iterative procedure is hereafter proposed. For a given number of ESSs, a heuristic based on voltage sensitivity analysis, which explicitly takes into account the network topology, is devised to select the most effective ESS location for voltage support. Then, the size of each storage unit is determined by solving a multi-period OPF problem. These steps are iterated for different numbers of ESSs, in order to minimize the overall configuration costs. Each step is illustrated separately in the following sections.

4.4.1 ESS Sizing

Assume that the set $\mathcal{S}$ is given, i.e., the ESS number and locations in the network have been decided. The considered storage sizing problem aims at finding the minimum total storage capacity making it possible to satisfy the voltage constraints (4.2) over a discrete time horizon $\mathcal{T}$. Typically, $\mathcal{T}$ spans one day or one week, in order to account for the periodicity of demand and generation profiles. Optimization variables in the sizing problem are the storage size $\overline{e}_s$ and the real and reactive power $p_s^{ess}(t-1)$ and $q_s^{ess}(t-1)$ exchanged by each storage unit, in addition to the typical OPF optimization variables. Conversely, demand and generation profiles are given, e.g., they are extracted from a historical data set. To define the cost function to be optimized, it is worth recalling that the major amount of losses in a power system (estimated around 70% of the total) is in distribution lines. This suggests that minimizing line losses should be a primary objective in the operation of distribution networks. Moreover, the battery size is another parameter to take into account while planning ESS investments. In view of these considerations, the following cost is introduced

$$J(C^{loss}, C^{cap}) = C^{loss} + \gamma_{cap} C^{cap}, \tag{4.21}$$

where $C^{cap} = \sum_{s\in\mathcal{S}} \overline{e}_s$ [kWh] is the total installed storage capacity, $C^{loss} = \frac{1}{|\mathcal{T}|} \sum_{t\in\mathcal{T}} \sum_{k\in\mathcal{N}} p_k(t)\Delta t$ [kWh] represents the average total line losses per sampling time, $|\mathcal{T}|$ denotes the cardinality of the set $\mathcal{T}$, and $\gamma_{cap} \geq 0$ is a weighting factor. Hence, the convex version of the AC OPF problem with storage dynamics can be cast as

Problem 6 Optimal ESS sizing problem.

$$\begin{cases} J^* = \min\limits_{W(t),\, p_s^{ess}(t-1),\, q_s^{ess}(t-1),\, \overline{e}_s} \quad J(C^{loss}, C^{cap}) & \\ \text{s.t.:} & \\ \mathrm{Tr}\big(Y_k^p W(t)\big) = p_k^{res}(t) - p_k^{load}(t) - p_k^{ess}(t-1) & \text{(a)} \\ \mathrm{Tr}\big(Y_k^q W(t)\big) = q_k^{res}(t) - q_k^{load}(t) - q_k^{ess}(t-1) & \text{(b)} \\ \underline{v}_k^2 \leq \mathrm{Tr}\big(M_k W(t)\big) \leq \overline{v}_k^2 & \text{(c)} \\ \mathrm{Tr}\big(Y_{ij}^p W(t)\big) \leq \overline{p}_{ij} & \text{(d)} \\ e_s(1) = e_s^{start} & \text{(e)} \\ e_s(t+1) = e_s(t) + p_s^{ess}(t-1)\Delta t & \text{(f)} \\ 0 \leq e_s(t) \leq \overline{e}_s & \text{(g)} \\ \underline{p}_s^{ess} \leq p_s^{ess}(t-1) \leq \overline{p}_s^{ess} & \text{(h)} \\ \underline{q}_s^{ess} \leq q_s^{ess}(t-1) \leq \overline{q}_s^{ess} & \text{(i)} \\ W(t) \succeq 0 & \text{(l)} \\ k \in \mathcal{N}^L, \quad (i,j) \in \mathcal{E}, \quad s \in \mathcal{S}, \quad t \in \mathcal{T} & \end{cases} . \tag{4.22}$$

In Problem 6, constraints (4.22a) and (4.22b) express the power balance equation (4.6) with the inclusion of ESS variables.[1] Clearly, $p_k^{ess}(t-1) = q_k^{ess}(t-1) = 0$ for $k \in \mathcal{N}^L \backslash \mathcal{S}$. Constraints (4.22c) and (4.22d) impose voltage and line capacity

[1] In power systems, measurements of average power are typically labeled a posteriori, i.e., the power ascribed at time t actually represents the average power between $t-1$ and t. On the other hand, the ESS control inputs holding between $t-1$ and t are denoted by $p_s^{ess}(t-1)$ and $q_s^{ess}(t-1)$, consistently with the fact that they are decided at time $t-1$.

limits (4.2) and (4.3), respectively, and constraints (4.22e)–(4.22i) model the storage dynamics (4.16)–(4.20). Notice that the conventional OPF problem (4.7) is a static optimization problem solved independently at each time step. Here, the addition of a storage charge/discharge dynamics allows optimization across time and gives way to a multi-period OPF. Moreover, (4.22) can be lifted to the original full non-linear multi-period OPF by simply adding the additional non-convex constraint rank $\big(\boldsymbol{W}(t)\big) = 1$ (see, e.g., [30, 31]).

Remark 5 Although Problem 6 is formulated in a deterministic framework, in a real context all variables are affected by uncertainty. These can be modeled by considering the data as realizations of suitable stochastic processes. This can be modeled by considering the data in each time horizon $\mathcal{T}$ as a sample of the cyclostationary stochastic process involving both renewables and loads. In this respect, the optimal cost J^*, as well as the optimal ESS sizes $\overline{e}^*$, are stochastic quantities, since the cost function (4.21) depends on the particular realization of demand and generation profiles over the considered time horizon. Data in each time frame $\mathcal{T}$ is considered as a sample of the cyclostationary process involving both renewable generation and loads. In order to guarantee feasibility of Problem 6, the size of the s-th device must be selected as the largest $\overline{e}^*$, taken with respect to all possible realizations of demand and generation profiles. When historical data are available, a viable approach is to solve Problem 6 separately for each sample in the data set, and then size each ESS according to the largest capacity found. Formally, let $\overline{e}_s^{*,d}$ denote the optimal size of storage unit s for the demand and generation profiles of sample $d = 1, \dots, D$. Then, the final size E_s^* of the s-th ESS is selected as

$$E_s^* = \max_{d=1,\dots,D} \overline{e}_s^{*,d}. \tag{4.23}$$

In this study, the time span of a sample will be assumed to be one day. In order to link coherently consecutive days, the additional constraint stating that the storage level at the beginning of a day equals that at the end of the day, namely

$$\sum_{t \in \mathcal{T}} p_s^{ess}(t-1) = 0, \quad \forall s \in \mathcal{S}, \tag{4.24}$$

is added in (4.22).

Remark 6 If needed, the ramp rate limits $\underline{p}_s^{ess}$ and $\overline{p}_s^{ess}$ can be made dependent on the ESS size $\overline{e}_s$ by replacing (4.18) and (4.22h) with

$$-\underline{\rho}_k \overline{e}_s \le p_s^{ess}(t-1) \le \overline{\rho}_k \overline{e}_s, \tag{4.25}$$

for fixed positive constants $\underline{\rho}_k$ and $\overline{\rho}_k$. Since (4.25) is convex, this extension does not change the computational complexity of the optimization problem (4.22).

4.4.2 ESS Siting

In this section, a procedure for finding a suitable placement of a given number of ESSs in a radial distribution network, based on Clustering and Sensitivity Analysis (CSA), is presented. The underlying idea consists in partitioning the network into n_c disjoint subnetworks, and then selecting the most appropriate bus within each subnetwork where to deploy an ESS. The CSA algorithm is presented below.

CSA algorithm

(i) *Network clustering*. First, the node set $\mathcal{N}^L$ is partitioned into n_c disjoint subsets. To this aim, a clustering algorithm is run on an auxiliary weighted graph, consisting of a complete graph built over $\mathcal{N}^L$. The weight associated to the edge (h, k) is equal to the (h, k)-entry of the *voltage sensitivity matrix* $\boldsymbol{\Psi} \in \mathbb{R}^{(n-1)\times(n-1)}$ (see, e.g., [35]):

$$\Psi_{hk} = \partial|V_k|/\partial p_h, \quad h, k \in \mathcal{N}^L. \tag{4.26}$$

The value Ψ_{hk} is a measure of how much the voltage at bus k is sensitive to active power injection at bus h. The effect of reactive power injection q_h on bus voltages is neglected since $\partial|V_k|/\partial q_h \ll \partial|V_k|/\partial p_h$ in LV networks. Among several methods for graph clustering available in the literature (see, e.g., [36]), the algorithm adopted in this procedure is based on [37]. It searches for n_c subgraphs which form a partition of the original graph, while minimizing the sum of the weights associated to the removed edges. In other words, this amounts to constructing the partition of the auxiliary graph in order to minimize the sum of the sensitivity values associated to the removed connections. This means that a pair of buses with high mutual sensitivity values are likely to end up in the same subset. The outcome of the clustering algorithm is a partition $\mathcal{N}_i, i = 1, \ldots, n_c$, of the node set $\mathcal{N}^L$, from which subnetworks $\Upsilon_i = (\mathcal{N}_i, \mathcal{E}_i)$ are reconstructed by defining $\mathcal{E}_i = \{(h_i, k_i) \in \mathcal{E} \, : \, h_i \in \mathcal{N}_i, k_i \in \mathcal{N}_i\}$.

(ii) *Candidate nodes*. For each subnetwork Υ_i, $i = 1, \ldots, n_c$, the *critical nodes* are identified as the nodes with generation and the nodes that are leaves of the original network. Then, the set Ω_i of *candidate nodes* is formed with all the nodes along the paths connecting any pair of critical nodes of Υ_i.

(iii) *Bus selection*. The criterion according to which the best node among all candidate nodes is selected exploits the voltage sensitivity (4.26) again. The ESS for subnetwork i is placed at the node

$$k_i^* = \arg\max_{k_i \in \Omega_i} \min_{h_i \in \mathcal{N}_i \setminus \{k_i\}} \Psi_{h_i k_i}. \tag{4.27}$$

This choice aims at maximizing the controllability of the voltage in the subnetwork.

The rationale behind the CSA algorithm is to exploit the voltage sensitivity matrix in order to determine the best ESS locations. Indeed, the voltage sensitivity matrix helps identify the most effective connections among nodes with the aim of maximizing the effect of power injection on voltage variation. Clearly, if a subnetwork Υ_i resulting from step (i) does not contain any node affected by voltage problems, Υ_i can be safely left without ESS, thus skipping steps (ii) and (iii) of the procedure for that subnetwork.

4.4.3 ESS Cardinality Selection

The parameter n_c, which is an upper bound to the final number of ESSs to be installed, is an input to the CSA algorithm, and must be selected by trading-off fixed and variable costs of storage configuration. A small number of ESSs results in smaller installation and maintenance costs, but typically requires a larger total capacity to be installed in order to face all possible network operating conditions with fewer ESSs. In this respect, a possible selection strategy for the number of subnetworks is to repeat the CSA algorithm for increasing values of n_c, find the optimal size of each device by solving Problem 6, and then evaluate the corresponding total configuration cost as

$$C^{conf}(n_c) = c_f n_s + c_v \mathbf{E}[J^*], \tag{4.28}$$

where n_s is the number of ESSs resulting from running the CSA algorithm with n_c clusters, J^* is the optimal cost of Problem 6, c_f [k€] accounts for the fixed costs related to a single device and c_v [€/kWh] is the unitary cost associated with J^*. Expectation $\mathbf{E}[\cdot]$ is taken with respect to demand and generation probability distributions. In fact, J^* is a random variable as explained in Remark 5. In practice, the expected value $\mathbf{E}[J^*]$ is replaced by the sample mean $\bar{J}^*$ computed over the available data set. Formally, let $J^{*,d}$ be the optimal cost of (4.22) for the demand and generation profiles of sample $d = 1, \ldots, D$, then

$$\bar{J}^* = \frac{1}{D} \sum_{d=1}^{D} J^{*,d}. \tag{4.29}$$

Remark 7 Since the voltage sensitivity matrix $\boldsymbol{\Psi}$ cannot be computed analytically, it is estimated numerically. This is done by evaluating the voltage variation at node k after a unit power injection at node h, which amounts to solving a load flow problem. While it is true that the entries of (4.26) do depend on the particular load and generation profiles, it turns out that their variation is negligible even when very different profiles are considered. This confirms that $\boldsymbol{\Psi}$ is determined mostly by the network topology and admittances of the lines, as already observed in previous works [9, 35]. By virtue of this observation, matrix $\boldsymbol{\Psi}$ is computed only once, by considering the average load and generation profiles. The same matrix is then used for solving points (i) and (iii) of the CSA procedure, irrespective of the number of clusters selected.

The overall ESS allocation procedure is summarized in Algorithm 2. The CSA algorithm for ESS siting is quite inexpensive. In fact, the hardest task of the CSA algorithm is the network clustering (step CSA.1), which is however carried out in few seconds for networks including up to a hundred buses (on a 2.4 GHz single-core CPU). The overall computational burden of the procedure is dictated by the solution of the sizing problem. Due to the time correlation introduced by the presence of ESSs, it is necessary to solve a multi-period OPF. Moreover, the full AC OPF need be solved since LV networks are considered. In this respect, the adopted SDP approximation yields a significant complexity reduction. Nevertheless, since the solution of the multi-period OPF is repeated for all the days in the historical data set, this task is by far the most time consuming one. Parallel implementation of this step could be effective in reducing the overall computation time.

Algorithm 2 Summary of the ESS allocation procedure

for $i = 1$ **to** $n - 1$ **do**
  $\mathcal{S}(i) \leftarrow$ run the CSA algorithm with $n_c = i$;
  for $d = 1$ **to** D **do**
    $(\overline{e}_s^{*,d}, J^{*,d}) \leftarrow$ solve Problem 6 with $\mathcal{S} = \mathcal{S}(i)$;
  end for
  $E_s^*(i) \leftarrow$ evaluate (4.23), $s \in \mathcal{S}(i)$;
  $C^{conf}(i) \leftarrow$ evaluate (4.28);
end for
$i^* = \arg\min_i C^{conf}(i)$;
place ESSs at $\mathcal{S}(i^*)$, with sizes $E_s^*(i^*)$.

4.5 ESS Operation

In this section, a control scheme based on a receding horizon approach for optimal operation of ESSs connected to the network is proposed. The idea is to use forecasts of demand and generation to "anticipate" possible voltage problems, and counteract them in advance. These forecasts are computed via simple statistical models requiring a limited set of real-time measurements, namely the power exchanged with the MV network, and meteorological variables (e.g., outdoor temperature and solar irradiance) provided by a local weather station. A simple procedure is presented to infer demand and generation at each bus of the network from aggregate forecasts at the secondary substation. Forecasts of demand and generation are used in a multi-period OPF, which provides the set points of active and reactive power to be exchanged by the installed ESSs. The objective function of the formulated optimization problem takes into account both the cost of storage use and line losses. The model predictive control paradigm is applied to voltage control also in [38, 39] for grid-connected microgrids, but without a proper representation of power flows and voltages, and in [40, 41], but for MV networks not equipped with ESS.

In the problem formulation, it is provisionally assumed that future demand and generation, i.e., the quantities $p_k^{load}(t)$, $q_k^{load}(t)$, $p_k^{res}(t)$ and $q_k^{res}(t)$ at buses $k \in \mathcal{N}^L$, are known. This assumption will be removed in the next section. At time t, the amount of active and reactive power exchanged by each ESS is computed by solving a suitable optimal control problem over the time horizon $\mathcal{T} = \{t+1, \ldots, t+H\}$. The parameter $H > 0$ represents the number of time periods included in the control horizon. The battery degradation, depending on both usage and energy level, is the main issue to take into account while operating an ESS. In view of these considerations, the following cost is introduced

$$C^{oper}(t) = C^{loss}(t) + \gamma_{ess} C^{ess}(t), \tag{4.30}$$

where $C^{loss}(t) = \sum_{k \in \mathcal{N}} p_k(t)\Delta t$ represents the total real losses in the network between $t-1$ and t, $C^{ess}(t) = \sum_{s \in \mathcal{S}} \left|p_s^{ess}(t-1)\right|\Delta t$ is a measure of the battery usage over the same time interval, and the term $\gamma_{ess} \geq 0$ is a suitable weight. At time t, the objective is to find an ESS control policy such that the sum of the costs (4.30) over the considered horizon $\mathcal{T}$ is minimized, while satisfying the voltage quality and line constraints. This translates into the following multi-period convex OPF problem.

Problem 7 Optimal ESS control problem.

$$\begin{cases} \displaystyle\min_{\boldsymbol{W}(\tau), p_s^{ess}(\tau-1), q_s^{ess}(\tau-1)} \sum_{\tau=t+1}^{t+H} C^{oper}(\tau) + \gamma_{soc} C^{soc,end}(t+H) & \\ \text{s.t.:} & \\ \mathrm{Tr}\big(\boldsymbol{Y}_k^p \boldsymbol{W}(\tau)\big) = p_k^{res}(\tau) - p_k^{load}(\tau) - p_k^{ess}(\tau-1) & \text{(a)} \\ \mathrm{Tr}\big(\boldsymbol{Y}_k^q \boldsymbol{W}(\tau)\big) = q_k^{res}(\tau) - q_k^{load}(\tau) - q_k^{ess}(\tau-1) & \text{(b)} \\ \underline{v}_k^2 \leq \mathrm{Tr}\big(\boldsymbol{M}_k \boldsymbol{W}(\tau)\big) \leq \overline{v}_k^2 & \text{(c)} \\ \mathrm{Tr}\big(\boldsymbol{Y}_{ij}^p \boldsymbol{W}(\tau)\big) \leq \overline{p}_{ij} & \text{(d)} \\ e_s(\tau) = e_s(\tau-1) + p_s^{ess}(\tau-1)\Delta t & \text{(e)} \\ 0 \leq e_s(\tau) \leq \overline{e}_s & \text{(f)} \\ \underline{p}_s^{ess} \leq p_s^{ess}(\tau-1) \leq \overline{p}_s^{ess} & \text{(g)} \\ \underline{q}_s^{ess} \leq q_s^{ess}(\tau-1) \leq \overline{q}_s^{ess} & \text{(h)} \\ \boldsymbol{W}(t) \succeq 0 & \text{(i)} \\ k \in \mathcal{N}^L, \quad s \in \mathcal{S}, \quad (i,j) \in \mathcal{E}, \quad \tau \in \mathcal{T} & \end{cases} \tag{4.31}$$

In Problem 7, the cost function includes an additional term $C^{soc,end}(t+H)$ which depends on the value of the storage level at the end of the control horizon. Parameter $\gamma_{soc} \geq 0$ is a suitable weight. The cost on the terminal energy storage state is defined as

$$C^{soc,end}(t) = \sum_{s \in \mathcal{S}} \left|e_s(t) - \delta\overline{e}_s\right|, \tag{4.32}$$

where $\delta \in [0, 1]$ represents the desired state of charge for each single ESS. The purpose of $C^{soc,end}$ is to drive the ESS state of charge at the end of the control horizon towards the target δ, compatibly with the voltage and power flow constraints. In the following, the values of the optimal control variables $p_s^{ess}(t)$ and $q_s^{ess}(t)$, solution of Problem 7, will be denoted by $p_s^{*,ess}(t)$ and $q_s^{*,ess}(t)$.

4.5.1 *Receding Horizon Implementation*

The implementation of Problem 7 in a real setting has to deal with the fact that future demand and generation are unknown at time t. For this reason, the values $p_k^{load}(t+h)$, $q_k^{load}(t+h)$, $p_k^{res}(t+h)$ and $q_k^{res}(t+h)$, are replaced in practice with their h-step ahead forecasts computed at time t, namely $\hat{p}_k^{load}(t+h|t)$, $\hat{q}_k^{load}(t+h|t)$, $\hat{p}_k^{res}(t+h|t)$ and $\hat{q}_k^{res}(t+h|t)$, $h = 1, \ldots, H$.

When tackling ESS operation via Problem 7, one would like to have the control horizon H as large as possible, because the larger H, the more efficient the ESS operating policy in terms of line losses and storage costs. On the other hand, the use of load and generation forecasts implies that the ESS operating policy obtained by solving Problem 7 will likely not be optimal under the true realization of demand and generation. More importantly, it could also fail in satisfying the voltage quality and real power flow constraints. These issues become more and more prominent as the control horizon H is increased, because prediction accuracy typically gets worse for large lead times. Therefore, a suitable trade-off should be found when selecting the control horizon in real applications. Values ranging from two to three hours are deemed to be a realistic compromise for LV networks under consideration.

One way to mitigate the effects of uncertainties such as inaccurate forecasts, is to apply the receding horizon approach which is typical of model predictive control [42]. Roughly speaking, at time t Problem 7 is solved based on the load and generation forecasts available at that time. Then, only the values $p_s^{*,ess}(t)$ and $q_s^{*,ess}(t)$ are applied. The same steps are repeated at the next time instants by exploiting the updated load and generation forecasts that become available. The complete receding horizon procedure is reported in Algorithm 3. The second step of Algorithm 3, requiring to compute demand and generation forecasts for each bus of the network, is described in detail in Sect. 4.5.2.

Algorithm 3 Receding horizon procedure for ESS operation

for $t = 0, 1, \ldots$ **do**
  Acquire the current battery state $e_s(t)$
  For each $k \in \mathcal{N}^L$, compute forecasts $\hat{p}_k^{load}(t+h|t)$, $\hat{q}_k^{load}(t+h|t)$, $\hat{p}_k^{res}(t+h|t)$ and $\hat{q}_k^{res}(t+h|t)$, $h = 1, \ldots, H$
  Solve Problem 7
  Apply control values $p_s^{*,ess}(t)$ and $q_s^{*,ess}(t)$
end for

4.5.2 Load and Generation Forecasting

The performance of the receding horizon approach for ESS operation described in Sect. 4.5.1 obviously depends on the accuracy of load and generation forecasts. In order to reliably predict load and generation at each bus of the network, measurements of these quantities would be needed in real-time, but unfortunately they are typically not available in LV networks due to the high cost of measurement and communication infrastructures. Many countries worldwide already adopt smart meters, but recorded measurements are typically transferred to data collectors in a batch way, e.g., monthly. Therefore, these measurements can be used for estimating the models of load and generation, but cannot be assumed to be available for the real-time network operation.

In this thesis, a measurement setting which is deemed to be as realistic as possible has been considered. Assume that a historical data set of load and generation profiles at all the buses of the LV network is available (e.g., from smart meters). This data set is used to estimate (and update, as soon as new data become available) the models of demand and generation in the network. To the purpose of ESS operation, and assuming that PV generation only is present, the requirement for real-time measurements is limited to:

- active power p_1 injected at the slack bus;
- solar irradiance I;
- outdoor temperature T.

In case generation from RES other than PV is present, measurements of additional meteorological variables could become necessary. To reduce the requirements for the communication infrastructure, solar irradiance and outdoor temperature are measured at the MV/LV substation, where the control logic is installed. This assumption is also practically motivated by the fact that LV networks typically do not cover large areas, and are therefore characterized by a limited variability of weather conditions.

The average active power p_1 injected at the slack bus between $t-1$ and t can be decomposed in the form

$$p_1(t) = p^{load}(t) - p^{res}(t) + p^{ess}(t), \tag{4.33}$$

where $p^{res}(t) = \sum_{k \in \mathcal{G}} p_k^{res}(t)$ is the aggregate generation of the network, $p^{ess}(t) = \sum_{s \in \mathcal{S}} p_s^{ess}(t-1)$ is the total active power exchanged by the ESSs, and $p^{load}(t) = \sum_{k \in \mathcal{N}^L} p_k^{load}(t) + p^{loss}(t)$ is the aggregate demand of the network, including the losses p^{loss} as a fictitious load. In the following the adopted methodologies to predict p^{load} and p^{res}, and infer from these predictions the forecasts of p_k^{load} and p_k^{res}, for all $k \in \mathcal{N}^L$, will be presented.

PV generation forecasting

For the sake of simplicity, it is assumed that generation in the LV network is only of PV type, as is often the case. Therefore, p^{res} in (4.33) can be seen as the aggregate

PV generation in the network and, following [43], it can be described as a function of the solar irradiance I and the outdoor temperature T via the PVUSA model

$$p^{res}(t) = I(t)\big[a + b\,I(t) + c\,T(t)\big], \tag{4.34}$$

where $a > 0$, $b < 0$ and $c < 0$ are the model parameters. The most attractive feature of the PVUSA model is that it is linear-in-the-parameters, and therefore parameter estimation can be performed very efficiently via least squares [44]. Here it is assumed that a, b and c were estimated from the available historical data set of generation, solar irradiance and outdoor temperature.

A strategy for predicting the aggregate generation p^{res} consists in substituting I and T in (4.34) with their forecasts:

$$\hat{p}^{res}(t+h|t) = \hat{I}(t+h|t)\big[a + b\,\hat{I}(t+h|t) + c\,\hat{T}(t+h|t)\big]. \tag{4.35}$$

Forecasting models of solar irradiance, mainly based on time series analysis and artificial neural networks, can be found in the literature, see, e.g., [45, 46]. A crucial point for solar irradiance forecasting is cloud covering, which represents the main source of reduction of solar radiation. Unfortunately, this phenomenon is characterized by a high variability in time and space, and is consequently very hard to predict. Currently, the cloud cover prediction is still a critical aspect of the meteorological science and only relatively few contribution can be fund in the literature (see, e.g., [47]). Modeling and prediction of outdoor temperature can be addressed through the combined use of exponential smoothing (ES) and autoregressive moving average (ARMA) models (e.g., see [48]). In the following, the h-step ahead predictor of I and T are denoted by f_h^I and f_h^T, respectively.

Finally, PV generation at each bus $k \in \mathcal{N}^L$ can be inferred from the predicted aggregate generation $\hat{p}^{res}$ as

$$\hat{p}_k^{res}(t+h|t) = \frac{\overline{p}_k^{res}}{\sum_{g\in\mathcal{G}} \overline{p}_g^{res}}\,\hat{p}^{res}(t+h|t), \tag{4.36}$$

where $\overline{p}_k^{res}$ denotes the installed PV power at bus k.

Load forecasting

Load demand exhibits strong seasonal behavior on different time scales. In particular, this applies to the aggregate demand p^{load} in (4.33), and therefore modeling and prediction of such a quantity can be addressed through ES-ARMA models (e.g., see [48]). With this in mind, the aggregate demand can be seen as a superposition of a seasonal/periodic pattern and a stochastic component. Therefore, p^{load} can be decomposed as

$$p^{load}(t) = p_b^{load}(t) + p_r^{load}(t), \tag{4.37}$$

where p_b^{load} is the seasonal component and p_r^{load} is the residual due to stochastic fluctuations. Based on (4.37), the problem of modeling p^{load} is split into the separate problems of modeling p_b^{load} and p_r^{load}. One way to obtain p_b^{load} is to apply an ES to p^{load} according to the smoothing equation

$$p_b^{load}(t) = \alpha\, p^{load}(t-\varrho) + (1-\alpha)\, p_b^{load}(t-\varrho), \tag{4.38}$$

where the delay ϱ depends on the periodicity of the time series, and $\alpha \in [0, 1]$ is the *smoothing parameter*. Values of α close to one give greater weight to recent observations, while small values of α determine a stronger smoothing of the stochastic fluctuations. The term p_r^{load} in (4.37) can be in principle modeled by resorting to different (possibly nonlinear) black-box model structures, but in many practical cases a simple ARMA model is sufficient. This corresponds to assume that p_r^{load} is generated by a mechanism of the type

$$p_r^{load}(t) = \frac{C(q)}{A(q)}\, \varepsilon(t), \tag{4.39}$$

where $\varepsilon(t)$ is a zero-mean i.i.d. stochastic process, and $A(q)$ and $C(q)$ are polynomials in the backward shift operator q^{-1}:

$$A(q) = 1 + a_1 q^{-1} + \ldots + a_{n_a} q^{-n_a} \tag{4.40a}$$

$$C(q) = 1 + c_1 q^{-1} + \ldots + c_{n_c} q^{-n_c}. \tag{4.40b}$$

The model combining (4.38) and (4.39) is referred to as ES-ARMA model. For fixed model orders n_a and n_c, the parameters of the ES-ARMA model (including the smoothing parameter α) are typically estimated by minimizing the mean square error (MSE) over estimation data. Order selection is performed by estimating optimal ES-ARMA models with different values of n_a and n_c, and then selecting the model with smallest MSE over validation data. When the ES-ARMA model is used for prediction, forecasts $\hat{p}^{load}(t+h|t)$ are computed according to the equation

$$\hat{p}^{load}(t+h|t) = p_b^{load}(t+h) + \hat{p}_r^{load}(t+h|t), \tag{4.41}$$

where $\hat{p}_r^{load}(t+h|t)$ is obtained by applying the linear minimum variance h-step ahead predictor derived from (4.39) [49]. To ease the notation, in the following the h-step ahead predictor of p^{load} is denote by f_h^{load}.

One practical issue to be considered when constructing the predictors f_h^{load}, is that measurements of p^{load} are not available in real-time, and hence pseudo-measurements $\widetilde{p}^{load}$ should be used in place of p^{load}. These can be defined from (4.33) as

$$\widetilde{p}^{load}(t) = p_1(t) + \widetilde{p}^{res}(t) - p^{ess}(t), \tag{4.42}$$

where $\widetilde{p}^{res}$ is the pseudo-measurement of p^{res} computed by substituting the measured values of I and T into (4.34). Note that, in (4.42), both p_1 and p^{ess} are known: the former is measured, while the latter is decided by the ESS control policy at time

---

Algorithm 4 Computation of demand and generation forecasts

Require: Measurements of $p_1(t)$, $p^{ess}(t)$, $I(t)$ and $T(t)$
  Compute $\widetilde{p}^{res}(t)$ from (4.34)
  Compute $\widetilde{p}^{load}(t)$ from (4.42)
  for $h = 1, \ldots, H$ **do**
    Compute $\hat{I}(t+h|t) = f_h^I(I(t), I(t-1), \ldots)$
    Compute $\hat{T}(t+h|t) = f_h^T(T(t), T(t-1), \ldots)$
    Compute $\hat{p}^{res}(t+h|t)$ from (4.35)
    Compute $\hat{p}^{load}(t+h|t) = f_h^{load}(\widetilde{p}^{load}(t), \widetilde{p}^{load}(t-1), \ldots)$
    for $k \in \mathcal{N}^L$ **do**
      Compute $\hat{p}_k^{res}(t+h|t)$ from (4.36)
      Compute $\hat{p}_k^{load}(t+h|t)$ from (4.44)
    end for
  end for

$t-1$ (see Sect. 4.5.1). Once a forecast of the aggregate demand p^{load} is available, the problem is how to use this value to infer the demand at each bus $k \in \mathcal{N}^L$. To this aim, the average fraction of the aggregate demand at bus k and time τ of the day, $\tau = 0, 1, \ldots, N_d - 1$, where N_d is the number of samples per day, is introduced and denoted by $\omega_k(\tau)$. The value of $\omega_k(\tau)$ is estimated from a historical data set as

$$\omega_k(\tau) = \frac{1}{D} \sum_{d=0}^{D-1} \frac{p_k^{load}(\tau + d\, N_d)}{p^{load}(\tau + d\, N_d)}, \tag{4.43}$$

where D is the number of days in the data set. Given the predicted aggregate demand $\hat{p}^{load}(t+h|t)$, the forecast $\hat{p}_k^{load}(t+h|t)$ of the demand at bus $k \in \mathcal{N}^L$ is obtained as

$$\hat{p}_k^{load}(t+h|t) = \omega_k\big((t+h) \bmod N_d\big)\hat{p}^{load}(t+h|t), \tag{4.44}$$

where mod is the modulo operation. It is worth pointing out that this is, for instance, the method actually used by the main Italian DSO for monitoring LV networks. In practice, it is also possible to make the fraction $\omega_k(\tau)$ depend on the day of the week in order to take into account different patterns of the demand, e.g., in weekdays and weekends. Notice that $\sum_{k\in\mathcal{N}^L} \omega_k(\tau) < 1$, in general, $1 - \sum_{k\in\mathcal{N}^L} \omega_k(\tau)$ being the average fraction of the active power dissipated in the lines.

The proposed procedure for predicting demand and generation at each bus of the network is summarized in Algorithm 4. Notice that the algorithm only refers to active power forecasts. Once forecasts of p_k^{load} and p_k^{res} are available for each bus $k \in \mathcal{N}^L$, reactive power q_k^{load} and q_k^{res} can be predicted by applying a fixed power factor, which can be estimated from data, and is typically between 0.9 and 1 for loads, and almost 1 for PV generators connected to the network through grid-tie inverters. The block scheme of the overall procedure for ESS operation, consisting of the two main steps of prediction and receding horizon control, is shown in Fig. 4.1.

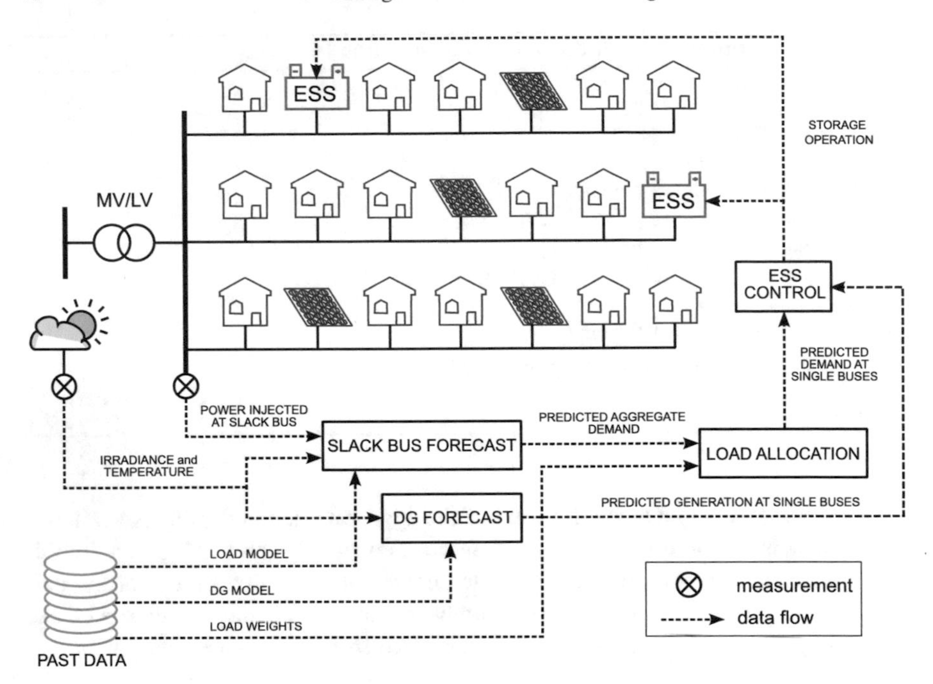

Fig. 4.1 Scheme of the proposed procedure for ESS operation. It consists of two main steps, namely prediction and control. At every sampling time, forecasts of both demand and generation are first computed for each bus of the network according to Algorithm 4. Then, according to Algorithm 3, the receding horizon control problem is solved, and the computed control inputs are applied to the ESSs

4.6 Experimental Results

Algorithms for the optimal ESS configuration and operation described so far are tested through several experiments. A real Italian LV networks, a modified version of the IEEE 34-bus test feeder, and 200 randomly generated radial networks are used for the validation of the CSA procedure. Then, the receding horizon algorithm for the optimal control of the installed ESSs is applied to real data from the Italian LV network.

Simulation setup

A portion of a real LV network, provided by the main Italian DSO (see Fig. 4.2) and denoted by IT-LV in the following, represents the main data set used for both planning and operation tests. The network consists of 17 buses, hosting 26 loads and 4 PV units. A total of 9 leaf and generation buses are present. Among them, only 8 critical nodes are retained since leaf node 3 does not experience any voltage problem. For all loads and generators, four months of active and reactive power profiles are

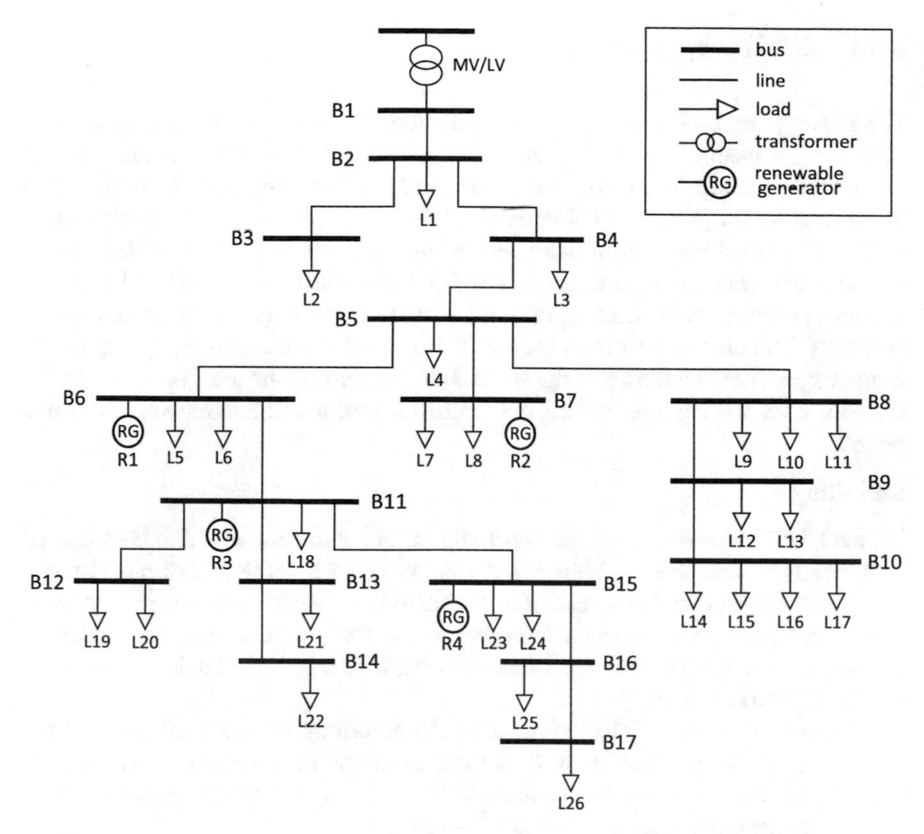

Fig. 4.2 IT-LV network

available with time step $\Delta t = 15$ min. These profiles are perturbed to originate both over- and undervoltages. This means that, in the absence of storage units installed in the network, voltage magnitudes violate the voltage quality constraints (4.2) at certain buses and time steps.

For all buses $k \in \mathcal{N}^L$, in accordance with the European Norm 50160, a 10% tolerance around the nominal voltage value is allowed in both directions, i.e., $\underline{v}_k = 0.9$ pu and $\overline{v}_k = 1.1$ pu in (4.2). The bounds $\overline{p}_{ij}$ in (4.3) are all set to 35 kW. The ESS technology adopted for the simulations features the following parameters. Active ramp limits in (4.18) are chosen such that $\overline{p}_s^{ess} = -\underline{p}_s^{ess} = 25$ kW, and the bounds $\underline{q}_s^{ess}$ and $\overline{q}_s^{ess}$ in (4.20) are set to keep the angle shift between -10 and $+10$ deg. All the results presented hereafter are obtained by using the CVX modelling toolbox [50] and the SeDuMi solver [51].

4.6.1 ESS Configuration

The strategy for ESS siting and sizing described in Sect. 4.4, is demonstrated on several experiments. The first experiment considers the IT-LV network described in the simulation setup. The second case study refers to a modified version of the IEEE 34-bus test feeder [52]. Two PV units are installed at buses 7 and 19, resulting in a total of 11 critical buses. Four daily demand and generation profiles, representative of four different operating conditions, are simulated with time step $\Delta t = 15$ min. In the third experiment, the CSA algorithm is tested on 200 randomly generated radial networks. The objective of the experiment is to evaluate the optimality gap of the solutions provided by the CSA algorithm. For this reason, the test is carried out for problem sizes making it possible to determine the optimal ESS siting via exhaustive search.

ESS siting

For the IT-LV network the voltage sensitivity matrix $\boldsymbol{\Psi}$ defined in (4.26) is computed numerically as described in Remark 7. A pseudocolor plot of $\boldsymbol{\Psi}$ is shown in Fig. 4.3. This kind of representation is useful to visualize the effect that a power injection at a given bus has on the voltage at the other buses. Cells with warmer colors denote strongly connected pairs of buses, whereas cells filled with colder colors correspond to weakly coupled buses.

For a given number of clusters n_c, the CSA algorithm returns a suitable number n_s of ESSs to be installed, as well as their locations in the network. The results obtained for the case $n_c = 6$ are shown in Fig. 4.4. The procedure terminates with $n_s = 2$ storage units, allocated at buses 7 and 11.

The CSA procedure is repeated for all possible values of n_c, ranging from 1 to 16 (see Table 4.1). Each column shows how nodes are grouped in clusters (labeled with

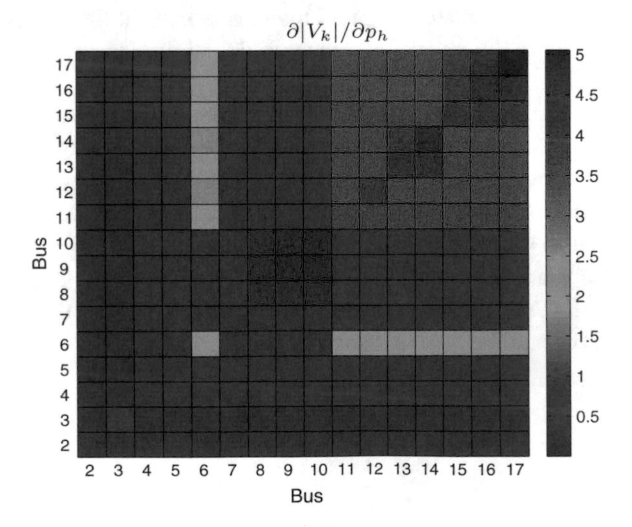

Fig. 4.3 Graphical representation of the voltage sensitivity matrix $\boldsymbol{\Psi}$ for the IT-LV network

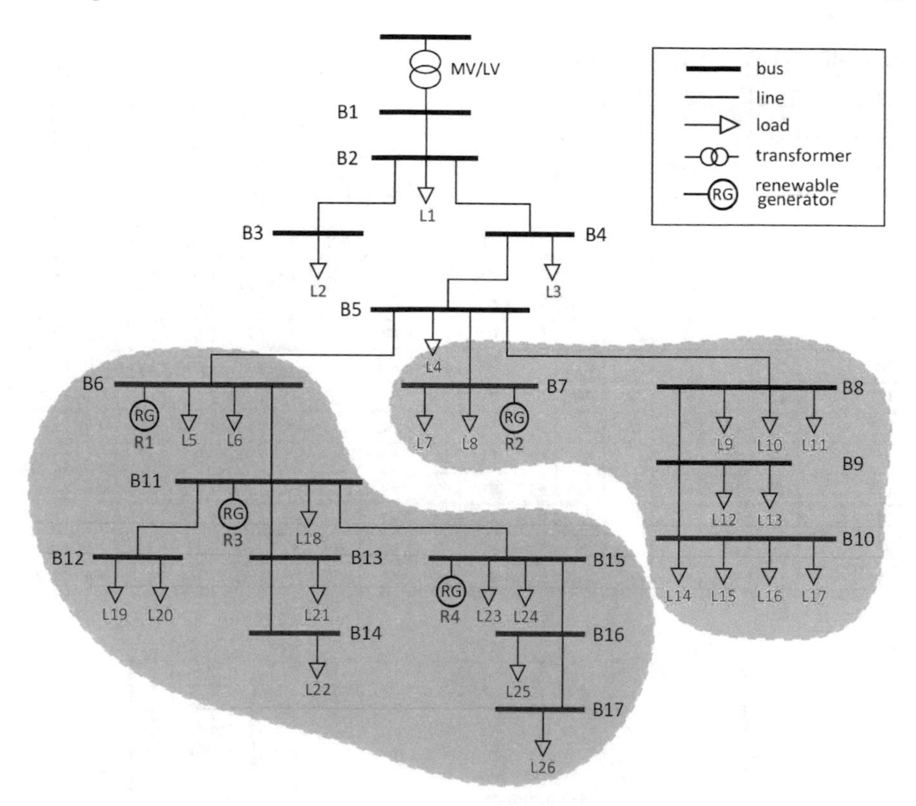

Fig. 4.4 Output of the CSA algorithm for $n_c = 6$. Two clusters (highlighted in green and yellow) contain a nonempty set of candidate nodes (red buses). The remaining four clusters are singletons, none of which is a candidate node

capital letters). Candidate buses are marked with a single vertical bar, while double vertical bars identify the nodes selected for hosting the ESSs according to (4.27). In the leftmost column, critical nodes are marked with an asterisk. By looking at the table, one can notice that irrespective of the chosen number of clusters, storage units are always allocated to a bus which is a critical node according to the definition given in Sect. 4.4.2. The number n_s of ESSs to be deployed as a function of the number of clusters n_c is shown in Fig. 4.5. It can be observed that n_s is typically strictly less than n_c. In the limit case in which each node makes up a different cluster (i.e., $n_c = 16$), the CSA algorithm returns $n_s = 8$, corresponding to the 8 critical buses present in the network. This confirms that such buses represent preferred locations in which to deploy ESSs, thus supporting the intuition which is at the basis of the CSA algorithm.

In order to assess the effectiveness of the siting procedure, the CSA algorithm is tested under four typical operating conditions of the original LV network (without ESSs):

Table 4.1 Output of the CSA algorithm for different number of clusters $n_c = 1, \ldots, 16$

bus\\n_c	1	2	3	4	5	6	7	8	9	10	11	12	13	14	15	16
2	A	A	A	A	A	A	A	A	A	A	A	A	A	A	A	A
3	A	B	B	B	B	B	B	B	B	B	B	B	B	B	B	B
4	A	B	C	C	C	C	C	C	C	C	C	C	C	C	C	C
5	$A\vert$	$B\vert$	$C\vert$	$D\vert$	D	D	D	D	D	D	D	D	D	D	D	D
6*	$A\vert$	$B\vert$	$C\vert$	$D\vert$	$E\vert$	$F\vert$	$F\vert$	$H\Vert$	$H\Vert$	$H\Vert$	$H\Vert$	$H\Vert$	$H\Vert$	$H\Vert$	$H\Vert$	$H\Vert$
7*	$A\vert$	$B\vert$	$C\vert$	$D\vert$	$E\Vert$	$E\Vert$	$G\Vert$	$G\Vert$	$G\Vert$	$G\Vert$	$G\Vert$	$G\Vert$	$G\Vert$	$G\Vert$	$G\Vert$	$G\Vert$
8	$A\Vert$	$B\Vert$	$C\Vert$	$D\Vert$	E	E	E	E	I	I	I	I	I	I	I	I
9	$A\vert$	$B\vert$	$C\vert$	$D\vert$	E	E	E	E	E	L	L	L	L	L	L	L
10*	$A\vert$	$B\vert$	$C\vert$	$D\vert$	$E\vert$	$E\vert$	$E\Vert$	$E\Vert$	$E\Vert$	$E\Vert$	$E\Vert$	$E\Vert$	$E\Vert$	$E\Vert$	$E\Vert$	$E\Vert$
11*	$A\vert$	$B\vert$	$C\vert$	$D\vert$	$E\vert$	$F\Vert$	$F\Vert$	$F\Vert$	$F\Vert$	$F\Vert$	$F\Vert$	$F\Vert$	$O\Vert$	$O\Vert$	$O\Vert$	$O\Vert$
12*	$A\vert$	$B\vert$	$C\vert$	$D\vert$	$E\vert$	$F\vert$	$F\vert$	$F\vert$	$F\vert$	$F\vert$	$M\Vert$	$M\Vert$	$M\Vert$	$M\Vert$	$M\Vert$	$M\Vert$
13	$A\vert$	$B\vert$	$C\vert$	$D\vert$	$E\vert$	$F\vert$	$F\vert$	$F\vert$	$F\vert$	$F\vert$	$F\vert$	N	N	P	P	P
14*	$A\vert$	$B\vert$	$C\vert$	$D\vert$	$E\vert$	$F\vert$	$F\vert$	$F\vert$	$F\vert$	$F\vert$	$F\vert$	$N\Vert$	$N\Vert$	$N\Vert$	$N\Vert$	$N\Vert$
15*	$A\vert$	$B\vert$	$C\vert$	$D\vert$	$E\vert$	$F\vert$	$F\vert$	$F\vert$	$F\vert$	$F\vert$	$F\vert$	$F\vert$	$F\Vert$	$F\Vert$	$Q\Vert$	$Q\Vert$
16	$A\vert$	$B\vert$	$C\vert$	$D\vert$	$E\vert$	$F\vert$	$F\vert$	$F\vert$	$F\vert$	$F\vert$	$F\vert$	$F\vert$	$F\vert$	$F\vert$	F	R
17*	$A\vert$	$B\vert$	$C\vert$	$D\vert$	$E\vert$	$F\vert$	$F\vert$	$F\vert$	$F\vert$	$F\vert$	$F\vert$	$F\vert$	$F\vert$	$F\vert$	$F\Vert$	$F\Vert$

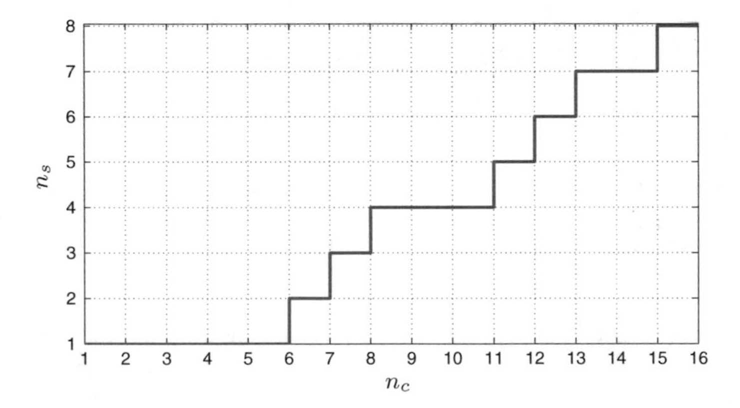

Fig. 4.5 Number of ESSs n_s returned by the CSA algorithm as a function of the number of clusters n_c for the IT-LV network

NV—neither over- nor undervoltage at any node;
OV—only overvoltage at some nodes;
UV—only undervoltage at some nodes;
UOV—both over- and undervoltage at some nodes.

For each of the previous scenarios, a representative day is extracted from the data set, and the corresponding load and generation profiles are considered. The CSA algorithm is run with $n_c = 6, 7, 8$, resulting in $n_s = 2, 3, 4$ storage units, respectively (see Fig. 4.5). Then, Problem 6 is solved by assuming ESS installed at the buses returned by the siting procedure. Denote by $J^{*,csa}$ the corresponding optimal cost in

(4.22). Such a quantity is then compared with the optimal cost $J^{*,opt}$, computed over all possible placements of n_s ESSs in the network (e.g., for $n_s = 2$, 120 different ESS placements have to be considered). The ratio $J^{*,csa}/J^{*,opt}$ is a measure of the performance degradation which is incurred when using the CSA algorithm for ESS siting (see Table 4.2). The increase in the cost ranges from zero (UOV scenario) to 5.7% (UV scenario and $n_s = 2$). This means that for the day when both over- and undervoltages are experienced, the storage locations returned by the CSA procedure are optimal. Conversely, for the day featuring only undervoltages, the CSA allocation of the two ESSs implies a cost $J^{*,csa}$ which is 5.7% higher than the optimal one $J^{*,opt}$. Notice, however, that the optimal ESS placement (i.e., the one yielding the optimal cost $J^{*,opt}$), being in general different from day to day, is not physically realizable. This means that, for fixed n_s, there does not exist any placement of n_s ESSs yielding the optimal cost $J^{*,opt}$ for all the four days. In this respect, the solution provided by the CSA algorithm turns out to be a good trade-off among different operating conditions and performs quite well in all scenarios. Moreover, the computation of the optimal placement requires an exhaustive search among all possible allocations of n_s storage units. This is a very time-consuming task which becomes rapidly intractable as n_s grows. On a 2.4 GHz single-core CPU, it takes about 16 h to compute the first row of Table 4.2, while it takes more than 10 days to compute the third one. A more extensive experimental validation of the CSA algorithm is carried out for the case $n_s = 2$. The same quantity reported in Table 4.2 is computed for all the 120 days available in the data set. The results, shown in Fig. 4.6a, confirm the good performance of the algorithm under every operating condition. Indeed, the few days for which the cost increase is above 10%, correspond to days experiencing undervoltages and for which ad hoc storage allocations result in a low optimal cost $J^{*,opt}$. However, the satisfactory behavior of the CSA algorithm on average can be appreciated by comparing the sample means $\bar{J}^{*,opt}$ and $\bar{J}^{*,csa}$, computed by averaging over the 120 days. Specifically, it is obtained $\bar{J}^{*,opt} = 151$ kWh, $\bar{J}^{*,csa} = 156$ kWh, resulting in an average performance degradation of 3.5%. Fig. 4.6b shows the daily cost increase (i.e., $\bar{J}^{*,\ell} - \bar{J}^{*,opt}$), averaged over the whole data set, for all possible allocations indexed by $\ell = 1, \ldots, 120$ of two ESSs in the network. It turns out that the solution provided by the CSA algorithm ranks #11, with an average cost very close to the best possible ESS allocation (namely, buses 10 and 11, for this network and data set).

Table 4.2 Performance evaluation of the CSA algorithm on the IT-LV network ($J^{*,csa}/J^{*,opt}$)

n_s	NV	OV	UV	UOV
2	1.008	1.022	1.057	1.000
3	1.001	1.038	1.000	1.000
4	1.006	1.000	1.008	1.000

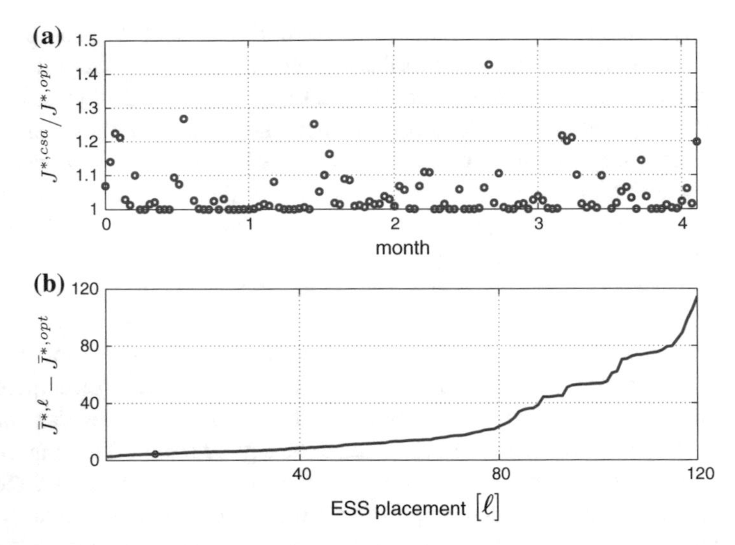

Fig. 4.6 Validation of the CSA algorithm for $n_s = 2$ for the IT-LV network. **a** Ratio $J^{*,csa}/J^{*,opt}$ for all the four months in the data set. **b** Average optimality gap $\bar{J}^{*,\ell} - \bar{J}^{*,opt}$ for all 120 possible placement of two ESSs in the network (sorted in increasing order). The ESS placement provided by the CSA algorithm ranks #11 (red dot)

ESS sizing

Once the CSA algorithm has returned the number and location of ESSs to be installed, the size of each storage unit is determined by solving Problem 6 for each day in the data set and then applying (4.23). An empty initial storage level (4.17) is assumed for all ESSs, i.e., $e_s^{start} = 0$ kWh. The weighting parameter in (4.21) is set to $\gamma_{cap} = 2.5 \cdot 10^{-3}$, whereas $c_f = 10$ k€ and $c_v = 575$ €/kWh are chosen in the total configuration cost (4.28).

An aspect to be considered is the quality of the solution found by solving the sizing problem. In fact, recall that (4.22) is a convex relaxation of the original non-convex problem, obtained by neglecting the rank-1 constraint. Feasibility and optimality of the solution of the relaxed problem for the original non-convex one are guaranteed if $\text{rank}\big(W(t)\big) = 1$ for all $t \in \mathcal{T}$. For the IT-LV network and the corresponding data set, the rank-1 constraint is never satisfied. Nevertheless, by solving a load flow problem with the ESS power exchange profiles $p_s^{ess}(t)$ and $q_s^{ess}(t)$ resulting from (4.22) with $\gamma_{cap} = 2.5 \cdot 10^{-3}$, the solution of the relaxed problem turns out to be always feasible for the original one, i.e., no violations of the constraints occur. Indeed, the ratio of the second to the first singular value of matrix $\boldsymbol{W}$, namely $\sigma_2(\boldsymbol{W})/\sigma_1(\boldsymbol{W})$, is of the order of 10^{-11} with the choice $\gamma_{cap} = 2.5 \cdot 10^{-3}$, i.e., matrix $\boldsymbol{W}$ is *almost* a rank-1 matrix.

The parameter γ_{cap} in (4.21) represents the weight assigned to the storage capacity term C^{cap} with respect to the line loss term C^{loss} in the cost function J. In order to tune it properly, several tests have been performed by considering load and generation

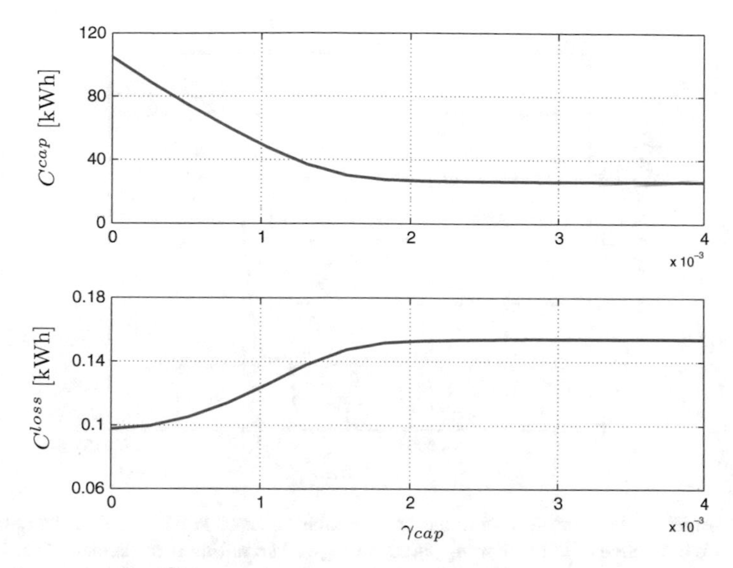

Fig. 4.7 Costs C^{cap} and C^{loss} at the optimum of Problem 6 as a function of the weighting parameter γ_{cap} for the IT-LV network. The set $\mathcal{S}$ is formed by ESSs placed at buses 7 and 11. Problem 6 is solved over one challenging day of the data set, i.e., a day featuring both over- and undervoltages in the absence of ESSs

profiles over a challenging day present in the data set, i.e., a day featuring both over- and undervoltages in the absence of ESSs. In Fig. 4.7, the optimal values of C^{cap} and C^{loss} as a function of γ_{cap} are shown. As expected, higher values of γ_{cap} yield solutions with smaller total installed ESS capacity, whereas smaller values of γ_{cap} result in smaller line losses. Moreover, it can be observed that as the weight γ_{cap} increases, C^{cap} cannot decrease below a lower bound in order to ensure feasibility of the voltage constraints in the optimization problem. Consequently, as C^{cap} tends to its lower bound, C^{loss} approaches a constant value as well. The choice $\gamma_{cap} = 2.5 \cdot 10^{-3}$ guarantees a total capacity C^{cap} as small as possible. Notice that γ_{cap} also affects the quality of the solution of the relaxed problem. For large values of γ_{cap}, it turns out that the solution of (4.22) is no longer feasible for the original non-convex problem. The reason for this can be understood by looking at the ratio $\sigma_2(\boldsymbol{W})/\sigma_1(\boldsymbol{W})$ in Fig. 4.8, which becomes non-negligible for $\gamma_{cap} > 9 \cdot 10^{-3}$.

ESS cardinality selection

The total configuration cost (4.28) as a function of the number of ESSs is shown in Fig. 4.9a. All the 120 days of the available data set are considered at this stage. For some ESS placements, there may exist days in the data set for which the sizing problem (4.22) does not admit a solution. For instance, this occurs when too few ESSs are deployed, or when ESSs are placed in nodes of the network providing little voltage support, no matter how big the storage size is. If this is the case, the day for which Problem 6 does not have a solution is termed *infeasible*. In practice, infeasible

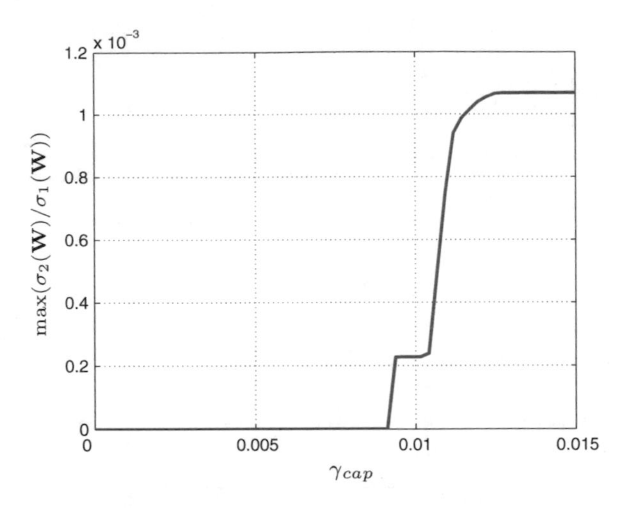

Fig. 4.8 Largest ratio of the second to the first singular value of $\mathbf{W}(t)$, $t \in \mathcal{T}$ at the optimum of Problem 6 as a function of the weighting parameter γ_{cap}. The setting is the same of Fig. 4.7

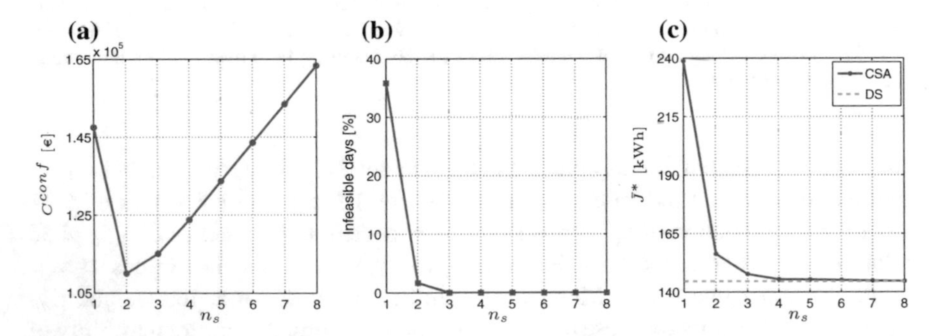

Fig. 4.9 Results of the proposed ESS configuration procedure for the IT-LV network. **a** Total configuration cost C^{conf} in (4.28), where the expected value $\mathbf{E}[J^*]$ is replaced with the average cost $\bar{J}^*$ in (4.29). **b** Percentage of infeasible days. **c** Average cost $\bar{J}^*$ (blue solid line) and lower bound provided by the distributed allocation strategy (red dashed line)

days are neglected when the average cost (4.29) and the final ESS sizes (4.23) are computed. However, the number of infeasible days provides useful information to make the final decision on the ESS allocation. The percentage of infeasible days as a function of n_s is shown in Fig. 4.9b. It is apparent that a suitable choice is $n_s = 2$, corresponding to the minimum total configuration cost $C^{conf} = 110$ k€, while the number of infeasible days is only 2 out 120 (less than 2%). Notice in Fig. 4.9a and b that increasing the number of ESSs from one to two leads to a consistent reduction of both the total configuration cost and the infeasible days. On the other hand, further increasing n_s would zero out the infeasible days, but at the price of a significant cost increase. The average cost $\bar{J}^*$, used in place of the expected value $\mathbf{E}[J^*]$ in (4.28), is shown in Fig. 4.9c. In particular, for $n_s = 2$, $\bar{J}^* = 156$ kWh (this is the value $\bar{J}^{*,csa}$

already reported in the previous analysis). Notice that $\bar{J}^*$ tends to a constant value as n_s increases. Recalling (4.28), this explains why C^{conf} in Fig. 4.9a grows linearly with n_s for $n_s \geq 4$.

Comparisons with alternative heuristics

In this section, the CSA algorithm is compared to other ESS allocation strategies.

The Distributed Strategy (DS) consists in solving Problem 6 for all the days in the data set with the assumption that a storage unit is available at each bus, i.e., $\mathcal{S} = \mathcal{N}^L$. Then, (4.23) is applied to determine the final ESS sizes. Since the optimal solution computed by DS is typically not sparse, i.e., $E_s^* \neq 0$ for most $s \in \mathcal{S}$, the number of allocated ESSs is high and hence the total cost C^{conf} is usually prohibitive (e.g., $C^{conf} = 243$ k€ for the IT-LV network under consideration). Nonetheless, considering DS is useful, since it provides a tight lower bound to (4.29), as can be observed in Fig. 4.9c. The same figure shows that increasing n_s from one to two allows one to fill about 85% of the gap between the value of $\bar{J}^*$ provided by the CSA algorithm for $n_s = 1$ and the lower bound provided by DS. Moreover, the choice $n_s = 2$ results in a total ESS size $C^{cap} = 64$ kWh which is slightly smaller than that provided by DS, namely 65 kWh (see Fig. 4.10, showing the sizes of the ESSs allocated at each bus by DS and by the CSA algorithm for $n_s = 2$).

The First Best Strategy (FBS) is based on DS. For a given number n_s of ESSs to be deployed, FBS builds the set $\mathcal{S}_{FBS}$ of buses with ESS by taking the n_s buses with the largest sizes E_s^* in the solution provided by DS. Then, the optimal size of each device is computed by solving Problem 6 with $\mathcal{S} = \mathcal{S}_{FBS}$ for all the days in the data set, and finally applying (4.23).

For $n_s = 2$, buses 6 and 15 are selected by FBS to host the ESSs (see Fig. 4.10). The total configuration cost is $C^{conf} = 111$ k€, with an average cost $\bar{J}^* = 157$ kWh. As can be observed, strategies FBS and the CSA algorithm achieve comparable results in terms of $\bar{J}^*$ and C^{conf} for $n_s = 2$. This occurs also for different values of the parameter

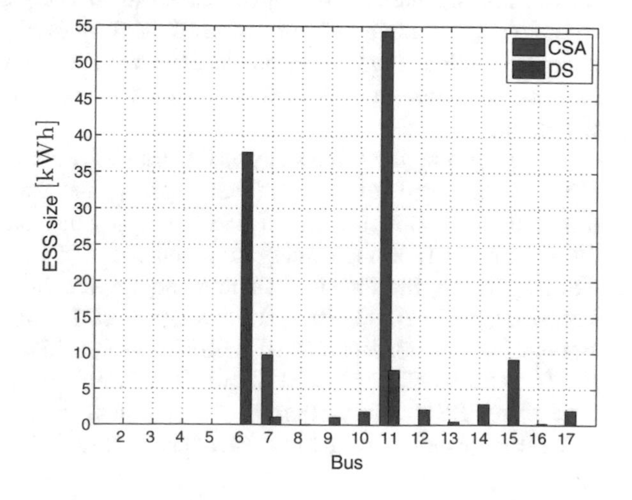

Fig. 4.10 Sizes E_s^* of the ESSs allocated by the distributed strategy (red) and by the CSA algorithm for $n_s = 2$ (blue) in the case of IT-LV network

Table 4.3 Performance evaluation of the CSA algorithm on the IEEE 34-bus test feeder ($J^{*,csa}/J^{*,opt}$)

n_s	NV	OV	UV	UOV
1	1.008	1.039	1.012	1.014
2	1.012	1.038	1.000	1.009
3	1.004	1.031	1.011	1.005

γ_{cap} in (4.21). However, FBS results in 11% of infeasible days for all $n_s \leq 6$. The reason for this lies in the fact that, as opposed to the CSA algoritm, FBS neglects the network topology, leaving the right-hand side of the network (see Fig. 4.2) without ESS support for $n_s \leq 6$. Under the considered demand and generation data set, the feeder composed of the buses {8, 9, 10} is likely to suffer from undervoltages. This cannot be solved with the solution provided by FBS until one chooses $n_s > 6$, thus introducing bus 10 in the set $\mathcal{S}_{FBS}$.

IEEE 34-bus case study

The ESS configuration procedure is validated on the IEEE 34-bus test feeder, by simulating demand and generation profiles of four days representative of the classes NV, OV, UV, and UOV, defined previously. As for the IT-LV network, the performance of the CSA algorithm is compared with the optimal solution, for different numbers of ESSs to be deployed. The results for the four classes are summarized in Table 4.3. The results confirm the effectiveness of the algorithm, with a maximum performance degradation smaller than 4%. With regard to computational aspects, for $n_s = 3$, the proposed procedure solves the siting and sizing problem for one day in approximately 18 min, almost totally required to solve Problem 6. As expected, the computation time taken by the CSA algorithm is of the order of seconds irrespective of the number of nodes, confirming that the siting procedure scales well with the network dimension. By contrast, the exhaustive search needed to compute the optimal ESS placement would require to solve 5456 problems of the form (4.22), implying an unacceptable computation time. For this reason, the exhaustive search needed for computing $J^{*,opt}$ is restricted to the set of the critical nodes, which are the favorite ones for the ESS allocation.

By performing the same analysis illustrated for the IT-LV network, the total cost C^{conf} is computed for different values of n_c, and using the four-day data set. In this case, the optimal number of ESSs to deploy into the IEEE 34-bus system turns out to be $n_s = 1$, corresponding to a total cost $C^{conf} = 52.8$ k€. According to the CSA algorithm, the location where to deploy the storage unit is bus 33, with an ESS size $C^{cap} = 20.9$ kWh, and an average cost function $\bar{J}^* = 74.1$ kWh. The FBS strategy for $n_s = 1$ selects bus 19 as ESS location, achieving comparable performance ($C^{conf} = 52.1$ k€, $C^{cap} = 21.3$ and $\bar{J}^* = 72.9$ kWh). However, the computation time of the FBS strategy, requiring the solution of an optimization problem assuming that a storage unit is available at each bus, is remarkably larger.

Test on random networks

To further assess the effectiveness of the allocation procedure, the CSA algorithm is run on 200 randomly generated radial networks, and results are compared with the optimal ones obtained via exhaustive search. All networks are characterized by the same node set $\mathcal{N}$ as the IT-LV network (bus 1 being always the slack bus), but differ in the edge set $\mathcal{E}$ and the admittance matrix $\boldsymbol{Y}$. The generation mechanism of the tree structure is recursive, starting from the slack bus as the root. Given a root node r and a set $\mathcal{L}$ of descendant nodes, m children c_i of r are randomly drawn from $\mathcal{L}$, where m ranges from 1 to 4 with equal probability. Then, the set of the remaining descendant nodes is randomly partitioned into m subsets $\mathcal{L}_i$. The same procedure is recursively repeated for each set $\mathcal{L}_i$ with root c_i, until the leaf nodes are reached. The depth of the 200 generated trees ranges from 3 to 7. Given the edge set $\mathcal{E}$, the admittance matrix $\boldsymbol{Y}$ is defined by generating randomly the distances between the nodes connected by an edge, and then multiplying the distances by a fixed admittance per unit length. On the other hand, demand and generation profiles are assumed to be a property of the buses (i.e., they are the same for all networks). The time step is 15 min, and the time horizon is one day, during which all the simulated networks feature over- or undervoltages at some buses.

For each network, the CSA algorithm is run with increasing values of n_c until it results in $n_s = 2$ deployed ESSs. Then, Problem 6 is solved by assuming the corresponding ESS placement. Let $J^{*,csa}$ be the optimal cost returned by (4.22) for that placement. Such a quantity is compared with the optimal cost $J^{*,opt}$, computed over all possible placements of two ESSs in the network. The empirical distribution of the ratio $J^{*,csa}/J^{*,opt}$ over the 200 networks is shown in Fig. 4.11a. It can be noticed that the solution provided by the CSA procedure is optimal in 45% of the cases, while only 9% of the cases reveal a degradation of the solution worse than 5%. Performance is even better when three ESSs are deployed, see Fig. 4.11b: the solution provided by the CSA procedure is optimal in 57% of the cases, while a

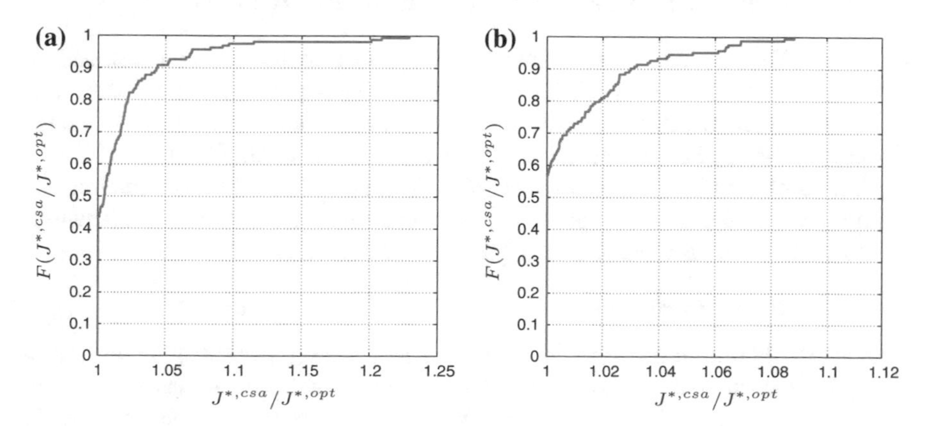

Fig. 4.11 Empirical distribution of the ratio $J^{*,csa}/J^{*,opt}$ over 200 random networks equipped with **a** $n_s = 2$, and **b** $n_s = 3$ ESSs

degradation of the solution worse than 5% occurs only in 5% of the cases, with a maximum degradation of 8%.

The results, obtained by applying the ESS allocation procedure to random topologies, confirm the soundness of the intuition underlying the CSA algorithm.

4.6.2 ESS Operation

The proposed ESS control algorithm is tested on the same data set described in the simulation setup. Since $\Delta t = 15$ min, the number of samples per day has been set accordingly, i.e., $N_d = 96$. Measurements of solar irradiance and outdoor temperature are also available with the same resolution. The historical data sets of demand and generation are used to estimate the load fractions $\omega_k(\tau)$ in (4.44) and the parameters of the PVUSA model (4.34), as well as of the ES-ARMA models of aggregate load, solar irradiance and outdoor temperature. The presence of two ESSs installed in the network is simulated. Location and size of ESSs are decided according to the CSA procedure. In particular, ESSs are deployed at buses 7 and 11, with capacities equal to $\overline{e}_7 = 10$ kWh and $\overline{e}_{11} = 54$ kWh, respectively (see Fig. 4.10). For both ESSs, the desired state of charge at the end of a control horizon is set to 50%, i.e., $\delta = 0.5$ in (4.32). This is done in order to ensure a suitable margin of operation in both directions (charge and discharge). Moreover, this choice aims to limit the depth of discharge, thus contributing to prolong battery life. Finally, the weights γ_{ess} in (4.30) and γ_{soc} in (4.31) are set to 0.015 and 96.5, respectively (different choices are discussed at the end of this section). Considering a control horizon of two hours, and therefore $H = 8$ in (4.31), Problem 7 is solved in few seconds for the 17-bus test network on a 3 GHz PC with 8 GB RAM. A check is made a posteriori to verify whether the optimal solution of the SDP problem is feasible for the original full non-linear OPF problem. In all the instances solved, the check provides a positive answer, thus certifying the optimality of the solution also for (4.31).

Remark 8 The fact that, in this application, the time step is 15 min (corresponding to the sampling time of the available data), while the computation of the control law requires few seconds, implies that the proposed approach is viable even for time steps closer to real-time. For instance, if data were available with sampling time of 5 min, the 2-h control horizon would include $H = 24$ time periods. In this situation, the control law could still be computed within a time step for the 17-bus test network, the computation time being less than 1 min on the 3 GHz PC.

Performance of ESS control

The control algorithm described in Sect. 4.5.1 is tested on a challenging week featuring both over- and undervoltages in the absence of ESSs. At each time step $t = 0, 1, 2, \ldots, N_w - 1$, where $N_w = 7N_d$ is the number of samples per week, demand and generation in the network are predicted for lead times h ranging from one sample (15 min ahead) to eight samples (2 h ahead). Then, the optimization problem (4.31) is solved with predicted demand and generation. The actual effect of the control inputs $p_s^{*,ess}(t)$ and $q_s^{*,ess}(t)$ is simulated by solving a load flow problem at time $t+1$ with true demand and generation.

In order to evaluate the benefits of using ESS control to tackle voltage problems, the average voltage violation at bus k over the considered time horizon is introduced as

$$\nu_k = \frac{1}{N_w} \sum_{t=0}^{N_w-1} [\underline{v}_k - |V_k(t)|]^+ + [|V_k(t)| - \overline{v}_k]^+, \tag{4.45}$$

where $[x]^+ = \max(x, 0)$. Let ν_k^{nc} be (4.45) evaluated in the absence of ESSs, and ν_k^{ess} in the case with ESSs. Then, the following performance index is defined

$$\eta = \left(1 - \frac{\sum_{k\in\mathcal{N}^L} \nu_k^{ess}}{\sum_{k\in\mathcal{N}^L} \nu_k^{nc}}\right) \cdot 100\%. \tag{4.46}$$

Assuming that the denominator of (4.46) is greater than zero (i.e., at least one violation of the voltage bounds occurs anywhere in the network over the considered week in the absence of ESSs), η quantifies how much the ESS control succeeds in preventing voltage problems. It turns out that $\eta = 99.2\%$, i.e., voltage violations are almost completely avoided or attenuated thanks to ESS control. More in detail, if one counts the number of time steps when the voltage violations occurring in the absence of ESSs are fully prevented by ESS control, the proposed algorithm reaches a performance of 96.6%. This means that only in 3.4% of the times a voltage violation is only partially mitigated. This can be observed in Fig. 4.12 (top), showing the plots of voltage magnitudes at all the buses of the network in the cases with and without ESSs. When ESS control is applied, only bus 6 still experiences light overvoltages in days 1, 3, 4 and 7, whereas all undervoltages are completely avoided over the whole week. From Fig. 4.12 (bottom), it can be also observed that ESS control reduces the total line losses, which stick to 65% of their nominal value (i.e., the value assumed in the case without ESSs). In particular, most of the reduction is achieved during peak hours of PV generation.

Figure 4.13 shows the plots of the storage levels $e_s(t)$ and the active power controls $p_s^{ess}(t)$ for both installed ESSs over the whole week. As expected, when no voltage violations are forecast, the state of charge of the ESSs is driven towards the desired value $\delta = 0.5$.

Finally, the effect of the weights γ_{ess} in (4.30) and γ_{soc} in (4.31) on the performance of the proposed receding horizon ESS control strategy is evaluated. Figure 4.14 shows the storage levels $e_s(t)$ over the whole week when the ESS controls are computed by solving Problem 7 with both weights doubled (top) and halved (bottom) with respect to the initial choice. In the first scenario, greater penalty is given to the use of storage, and consequently the available ESS capacity is only partially utilized with respect to the case of Fig. 4.13 (top). Moreover, the total line losses increase to 73% of their nominal value. In the second scenario, line losses receive greater penalty, and actually they further decrease to 60% of their nominal value. This is obtained at the expense

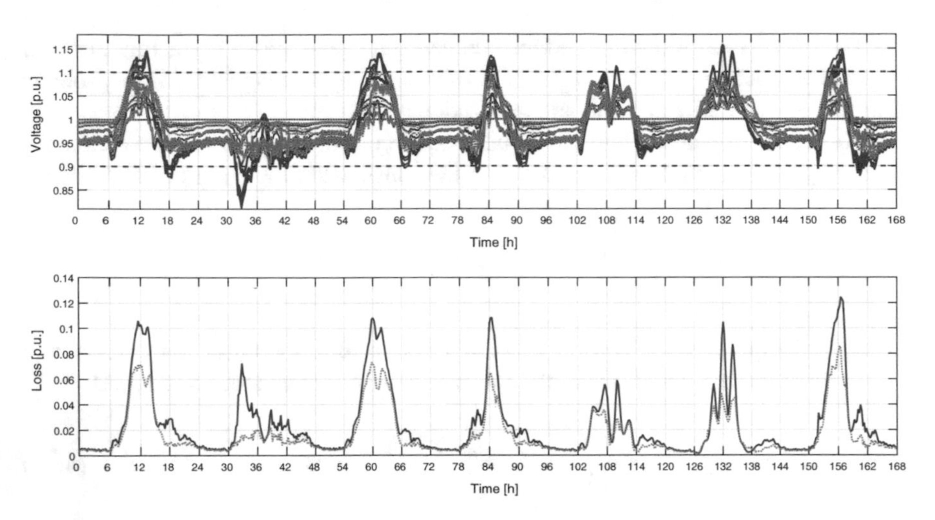

Fig. 4.12 Performance of the receding horizon ESS control algorithm. Top: Voltage limits (red dashed lines) and voltage magnitudes at all buses $k \in \mathcal{N}^L$, with ESSs (green dotted lines) and without ESSs (blue solid lines). Bottom: Line losses with ESSs (green dashed line) and without ESSs (blue solid line)

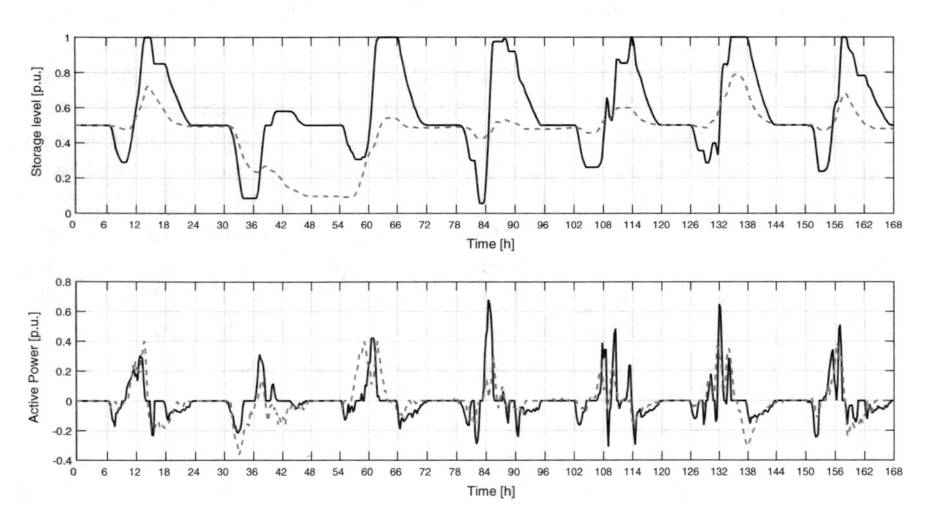

Fig. 4.13 Operation of ESSs at bus 7 (black solid line) and bus 11 (blue dashed line). Top: Storage levels $e_s(t)$. Bottom: Active power controls $r_s(t)$

of a greater use of storage, as can be noticed by comparing Fig. 4.14 (bottom) with Fig. 4.13 (top). However, even though the ESSs are operated in such different ways, performance in terms of voltage support is comparable. It turns out that $\eta = 97.3\%$ and $\eta = 99.5\%$ for the first and second scenario, respectively.

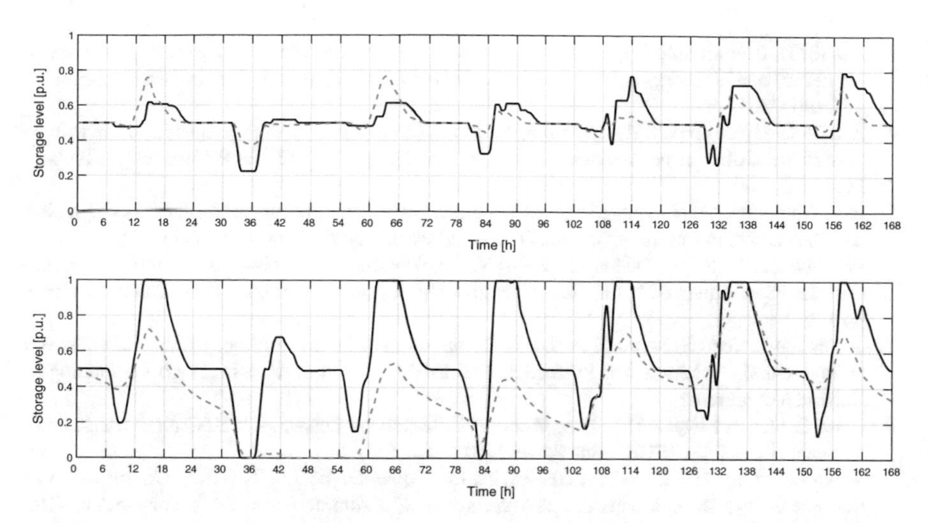

Fig. 4.14 Comparison of the storage levels $e_s(t)$ for different weights in the cost function of (4.31). Top: $\gamma_{ess} = 0.030$, $\gamma_{soc} = 193.00$. Bottom: $\gamma_{ess} = 0.0075$, $\gamma_{soc} = 48.25$

4.7 Conclusions

In this chapter, a new method for designing the number, location and size of ESSs in LV networks, is proposed. In order to reduce the computational burden required by the solution of the resulting optimization problem, a heuristic procedure is devised. A voltage sensitivity analysis has been proposed to circumvent the combinatorial nature of the siting problem and a multi-period OPF has been adopted to determine the size of each storage unit. In addition, the problem of controlling distributed storage devices has been addressed. The proposed algorithm, which is inspired by the classical receding horizon control approach, aims at minimizing line losses and the storage usage, while maintaining the voltages within the specified limits. Only the power exchanged at the slack bus and meteorological variables at the secondary substation are monitored in order to predict the future state of the network and any potential voltage rise/drop. The simulations show interesting results in terms of computational time and solution optimality.

References

1. Giannitrapani A, Paoletti S, Vicino A, Zarrilli D (2015) Algorithms for placement and sizing of energy storage systems in LV networks. In: Proceedings of IEEE conference on decision and control, pp 3945–3950
2. Giannitrapani A, Paoletti S, Vicino A, Zarrilli D (2017) Optimal allocation of energy storage systems for voltage control in LV distribution networks. IEEE Trans Smart Grid 8(6):2859–2870

3. Zarrilli D, Giannitrapani A, Paoletti S, Vicino A (2016) Receding horizon voltage control in LV networks with energy storage. In: Proceedings of IEEE conference on decision and control, pp 7496–7501
4. Zarrilli D, Giannitrapani A, Paoletti S, Vicino A (2018) Energy storage operation for voltage control in distribution networks: a receding horizon approach. IEEE Trans Control Syst Technol 26(2):599–609
5. Xu T, Taylor PC (2008) Voltage control techniques for electrical distribution networks including distributed generation. In: Proceedings of IFAC world congress, pp 11,967–11,971
6. Procopiou AT, Long C, Ochoa LF (2014) Voltage control in LV networks: an initial investigation. In: Proceedings of IEEE PES innovative smart grid technologies European conference, pp 1–6
7. Tuominen J, Repo S, Kulmala A (2014) Comparison of the low voltage distribution network voltage control schemes. In: Proceedings of IEEE PES innovative smart grid technologies European conference
8. Long C, Ochoa LF (2015) Voltage control of PV-rich LV networks: OLTC-fitted transformer and capacitor banks. IEEE Trans Power Syst 1–10
9. Wang L, Liang DH, Crossland AF, Taylor PC, Jones D, Wade NS (2015) Coordination of multiple energy storage units in a low-voltage distribution network. IEEE Trans Smart Grid 6(6):2906–2918
10. Nazaripouya H, Wang Y, Chu P, Pota HR, Gadh R (2015) Optimal sizing and placement of battery energy storage in distribution system based on solar size for voltage regulation. In: Proceedings of IEEE PES general meeting, pp 1–5
11. Nazaripouya H, Wang Y, Chu P, Pota HR, Gadh R (2015) Optimal siting and sizing of distributed energy storage systems via alternating direction method of multipliers. Int J Electr Power Energy Syst 72:33–39
12. U.S. Department of Energy (2016) Grid energy storage. http://energy.gov/oe/downloads/grid-energy-storage-december-2013. Accessed 27 Apr 2016
13. Int. Electrotechnical Commission (2016) Electrical energy storage. http://www.iec.ch/whitepaper/pdf/iecWP-energystorage-LR-en.pdf. Accessed 27 Apr 2016
14. Zarrilli D, Vicino A, Mancarella P (2016) Sharing energy resources in distribution networks: an initial investigation through OPF studies. In: Proceedings of IEEE PES innovative smart grid technologies asian conference, pp 306–311
15. Saint-Pierre A, Mancarella P Active distribution system management: a dual-horizon scheduling framework for dso/tso interface under uncertainty. IEEE Trans Smart Grid, to appear
16. Pandzic H, Wang Y, Qiu T, Dvorkin Y, Kirschen DS (2015) Near-optimal method for siting and sizing of distributed storage in a transmission network. IEEE Trans Power Syst 30(5):2288–2300
17. Zidar M, Georgilakis PS, Hatziargyriou ND, Capuder T, Škrlec D (2016) Review of energy storage allocation in power distribution networks: applications, methods and future research. IET Gener Transm Distrib 10(3):645–652
18. Hoffman M, Kintner-Meyer M, DeSteese J, Sadovsky A (2011) Analysis tools for sizing and placement of energy storage in grid applications. In: Proceedings of ASME international conference on energy sustainability, pp 1565–1573
19. Bucciarelli M, Giannitrapani A, Paoletti S, Vicino A, Zarrilli D (2016) Energy storage sizing for voltage control in LV networks under uncertainty on PV generation. In: Proceedings of IEEE international forum on research and technologies for society and industry leveraging a better tomorrow, pp 1–6
20. Zhao H, Wu Q, Huang S, Guo Q, Sun H, Xue Y (2015) Optimal siting and sizing of energy storage system for power systems with large-scale wind power integration. In: Proceedings of IEEE PowerTech Eindhoven, pp 1–6
21. Thrampoulidis C, Bose S, Hassibi B (2016) Optimal placement of distributed energy storage in power networks. IEEE Trans Autom Control 61(2):416–429
22. Wogrin S, Gayme DF (2015) Optimizing storage siting, sizing, and technology portfolios in transmission-constrained networks. IEEE Trans Power Syst 30(6):3304–3313

23. Ghofrani M, Arabali A, Etezadi-Amoli M, Fadali MS (2013) A framework for optimal placement of energy storage units within a power system with high wind penetration. IEEE Trans Sustain Energy 4(2):434–442
24. Taylor JA, Hover FS (2012) Convex models of distribution system reconfiguration. IEEE Trans Power Syst 27(3):1407–1413
25. Nick M, Cherkaoui R, Paolone M (2014) Optimal allocation of dispersed energy storage systems in active distribution networks for energy balance and grid support. IEEE Trans Power Syst 29(5):2300–2310
26. Bose S, Gayme DF, Topcu U, Chandy KM (2012) Optimal placement of energy storage in the grid. In: Proceedings of IEEE conference on decision and control, pp 5605–5612
27. Gayme D, Topcu U (2013) Optimal power flow with large-scale storage integration. IEEE Trans Power Syst 28(2):709–717
28. Torchio M, Magni L, Raimondo D (2015) A mixed integer SDP approach for the optimal placement of energy storage devices in power grids with renewable penetration. In: Proceedings of American control conference, pp 3892–3897
29. Torchio M, Magni L, Raimondo D (2015) On the effects of monitoring and control settings on voltage control in PV-rich LV networks. In: Proceedings of IEEE PES general meeting, pp 1–5
30. Lavaei J, Low SH (2012) Zero duality gap in optimal power flow problem. IEEE Trans Power Syst 27(1):92–107
31. Low SH (2014) Convex relaxation of optimal power flow-Part I: formulations and equivalence. IEEE Trans Control Netw Syst 1(1):15–27
32. Gan L, Low SH (2014) Convex relaxations and linear approximation for optimal power flow in multiphase radial networks. In: Proceedings of power systems computation conference, pp 1–9
33. Gan L, Low SH (2014) Convex relaxation of optimal power flow-Part II: exactness. IEEE Trans Control Netw Syst 1(2):177–189
34. Jabr RA, Karaki S, Korbane JA (2015) Robust multi-period OPF with storage and renewables. IEEE Trans Power Syst 30(5):2790–2799
35. Brenna M, De Berardinis E, Delli Carpini L, Foiadelli F, Paulon P, Petroni P, Sapienza G, Scrosati G, Zaninelli D (2013) Automatic distributed voltage control algorithm in smart grids applications. IEEE Trans Smart Grid 4(2):877–885
36. Nascimento MC, De Carvalho AC (2011) Spectral methods for graph clustering-a survey. Eur J Oper Res 211(2):221–231
37. Hespanha JP (2016) An efficient MATLAB algorithm for graph partitioning. http://www.ece.ucsb.edu/~hespanha/published/tr-ell-gp.pdf. Accessed 27 Apr 2016
38. Malysz P, Sirouspour S, Emadi A (2014) An optimal energy storage control strategy for grid-connected microgrids. IEEE Trans Smart Grid 5(4):1785–1796
39. Parisio A, Rikos E, Glielmo L (2014) A model predictive control approach to microgrid operation optimization. IEEE Trans Control Syst Technol 22(5):1813–1827
40. Valverde G, Van Cutsem T (2013) Model predictive control of voltages in active distribution networks. IEEE Trans Smart Grid 4(4):2152–2161
41. Farina M, Guagliardi A, Mariani F, Sandroni C, Scattolini R (2015) Model predictive control of voltage profiles in MV networks with distributed generation. Control Eng Pract 34:18–29
42. Maciejowski JM (2002) Predictive control with constraints. Pearson Education
43. Dows R, Gough E (1995) PVUSA procurement, acceptance, and rating practices for photovoltaic power plants. Pacific Gas and Electric Company, San Ramon, CA, Technical Report
44. Bianchini G, Paoletti S, Vicino A, Corti F, Nebiacolombo F (2013) Model estimation of photovoltaic power generation using partial information. In: Proceedings of IEEE PES innovative smart grid technologies European conference
45. Reikard G (2009) Predicting solar radiation at high resolutions: a comparison of time series forecasts. Sol Energy 83:342–349
46. Wu J, Chan C (2011) Prediction of hourly solar radiation using a novel hybrid model. Sol Energy 85:808–817

47. Pepe D, Bianchini G, Vicino A (2016) Model estimation for PV generation forecasting using cloud cover information. In: Proceedings of IEEE international energy conference, pp 1–6
48. Garulli A, Paoletti S, Vicino A (2015) Models and techniques for electric load forecasting in the presence of demand response. IEEE Trans Control Syst Technol 23(3):1087–1097
49. Box GEP, Jenkins GM, Reinsel GC (2008) Time series analysis: forecasting and control, 4th edn. Wiley
50. Grant M, Boyd S (2016) CVX: Matlab software for disciplined convex programming, version 2.1. http://cvxr.com/cvx. Accessed 27 Apr 2016
51. Sturm J (1999) Using SeDuMi 1.02, A Matlab toolbox for optimization over symmetric cones. Optim Methods Softw 11(1–4):625–653
52. Kersting W (1991) Radial distribution test feeders. IEEE Trans Power Syst 6(3):975–985

Chapter 5
Conclusions and Future Research

This final chapter contains a summary of the thesis contributions and a brief discussion of the achieved results and future research directions.

5.1 Summary of Contributions

The contributions of this thesis can be divided into three categories: bidding strategies for WPPs, DR integration in smart buildings, and ESS applications for voltage support in distribution networks.

5.1.1 Bidding Strategy for WPPs

The problem of offering renewable power in the electricity market featuring soft penalties, i.e. penalties that are applied only when the delivered power deviates from the nominal bid more than a given relative tolerance, has been addressed for WPPs. The optimal bidding strategy, based on the knowledge of the prior wind power statistics, has been derived analytically by maximizing the expected profit of the producer. Moreover, the use of additional knowledge, represented by wind speed forecasts provided by a meteorological service, to make more reliable bids, has been embedded in the bidding process by means of machine learning-based classification approaches. The performance of the optimal bidding strategy, both with and without classification, has been demonstrated on experimental data from a real Italian wind farm, and compared with that of the naive bidding strategy based on offering wind power forecasts computed by plugging the wind speed forecasts into the wind plant power curve. It has been proved that exploiting wind speed forecasts

D. Zarrilli, *Integration of Low Carbon Technologies in Smart Grids*,
Springer Theses, https://doi.org/10.1007/978-3-319-98358-5_5

through classification approaches allows one to improve consistently the profit of renewable power producers.

5.1.2 DR Integration in Smart Buildings

An innovative home heating management system under the hypothesis that the building participates in DR programs, has been proposed. A receding horizon control approach has been adopted for minimization of the energy bill, by exploiting a simplified version of the building model, while maintaining the thermal comfort of the occupants. Since the resulting optimization problem is a mixed integer linear programming problem which turns out to be manageable only for buildings with very few zones, a heuristic has been devised to make the algorithm applicable to realistic size cases. The derived control law has been tested on the close-to-reality simulator EnergyPlus to evaluate pros and cons of the presented algorithm. Numerical results show that the developed control approach involves almost negligible loss of accuracy with respect to the exact optimal solution.

5.1.3 ESS Applications for Voltage Support in Distribution Networks

The optimal configuration and operation of distributed ESSs for providing voltage support in low voltage networks has been investigated. In the planning stage the objective is to find the allocation which minimizes the total cost of storage devices, which depends both on the number of ESSs and on their size. Since the exact problem turns out to be intractable in realistic applications, a heuristics based on the voltage sensitivity analysis has been devised for circumventing the combinatorial nature of the siting problem. A real time control scheme based on a receding horizon approach to optimally operate ESSs installed in the network has been also devised. For the operation stage, the objective translates to minimizing both ESS use and line losses. The essential feature of the control approach lies in the very limited information needed to predict possible voltage problems, and to compute the ESS policy making it possible to counteract them in advance. The overall procedure has been tested on real data from an Italian LV network. Moreover, a modified version of the IEEE 34-bus test feeder and 200 randomly generated radial networks have been used for a further validation of the planning step. Simulations show that the overall procedure is amenable to fast and robust computation, while greatly enhancing the voltage quality at customers' premises.

5.2 Future Research Directions

Modern electricity power networks are undergoing dramatic changes, both at transmission and distribution level. They have to cope with increasing penetration of distributed and renewable energy sources, large data flows from intelligent power equipments and smart meters, and market pressure to run efficiently. This results in heavier work and stress for network/market operators. The market and energy management tools presented in this thesis provide effective answers to some of the aforementioned new challenges. However, the results obtained are by no means exhaustive and there are many aspects of the considered problems that still remain to be investigated.

Strategies for renewable power producers selling wind power in the electricity spot market have been developed in Chap. 2. Although the potential of the proposed bidding mechanism has been proved for the day-ahead market structure, it does not apply directly to different model setups, such as intraday or future markets. Moreover, it has been assumed that the electricity prices are known and given in advance. A suitable way to incorporate physical transmission capacity constraints and their effects on market price/penalties in the bidding process would be a valuable add-on to the market model presented in this thesis.

An appropriate heuristic based on the MPC approach has been devised in Chap. 3 for smart buildings participating in DR programs. An underlying assumption in the model is that the home energy management system only works for heating purposes. From both the theoretical and practical viewpoint, it would be interesting to see how the developed control algorithm can be extended to a more complex setup where the building is considered as a microgrid, including the entire HVAC plant providing heating and cooling services, different renewable generation sources, electric and thermal storage devices, electric appliances and other kinds of electric loads, such as electric vehicles.

Integration of ESSs in the power system is a key component of future smart grids. In Chap. 4 a novel approach was proposed to optimally deploy and operate such devices in radial distribution networks. An important aspect for future investigation is the possibility to extend the siting heuristic developed in the configuration step, which is strictly based on the physical properties of lines and on the topology of the network, to meshed grids. A further research direction could also investigate the case in which ESSs are considered a common property of the network, owned and managed by the whole community. Since economics is key to ESS applications, the possibility to fully exploit the ESS potential for providing multiple services (not only voltage support, but also frequency control, peak shaving, spinning reserves, etc.) could play a significant role for the successful advancement and use of this technology. In this respect, developing innovative ESS business models aiming at handling different networks issues is expected to be a very fruitful line of investigation.

Author Biography

Dr. Donato Zarrilli was born in Calitri, Italy, in 1988. He is currently a production supervisor and WCM specialist at Maserati S.p.A., in Turin, Italy.

He received the Laurea degree in engineering management (Honors) in 2012, and the Ph.D. degree in engineering and science of information (Honors) in 2016, both from the Università di Siena, Siena, Italy. During his Ph.D., he was also a visiting scientist at the School of Electrical and Electronic Engineering, University of Manchester, Manchester, U.K., in 2016. His research interests included modeling, forecasting, control and optimization techniques for the management of smart electrical grids. He has published over 15 journal articles and peer-reviewed conference papers, and has given talk and invited seminars at several workshops and international conferences. As a result of his effort, sincere attitude and passion to the research, he was awarded in 2017 the Prize as the best Italian Ph.D. thesis in control and system engineering by the Italian Society for Research in Systems and Control Engineering (SIDRA).

D. Zarrilli, *Integration of Low Carbon Technologies in Smart Grids*,
Springer Theses, https://doi.org/10.1007/978-3-319-98358-5

Printed in the United States
By Bookmasters